CAMBRIDGE MONOGRAPHS IN
EXPERIMENTAL BIOLOGY
No. 13

EDITORS:
T. A. BENNET-CLARK
GEORGE SALT (*General Editor*)
V. B. WIGGLESWORTH

# THE PHYSIOLOGY OF
# DIURNAL RHYTHMS

## THE SERIES

# THE PHYSIOLOGY OF
# DIURNAL RHYTHMS

BY

## JANET E. HARKER

*Fellow of Girton College*
*Cambridge*

## CAMBRIDGE
## AT THE UNIVERSITY PRESS
1964

PUBLISHED BY

THE SYNDICS OF THE CAMBRIDGE UNIVERSITY PRESS

Bentley House, 200 Euston Road, London, N.W. 1
American Branch: 32 East 57th Street, New York 22, N.Y.
West African Office: P.O. Box 33, Ibadan, Nigeria

©

CAMBRIDGE UNIVERSITY PRESS
1964

*Printed in Great Britain by Spottiswoode, Ballantyne & Co. Ltd.,
London and Colchester*

# CONTENTS

# ACKNOWLEDGEMENTS

I AM indebted to Dr E. A. C. MacRobbie for many stimulating discussions on the problems of diurnal rhythms and on the possible biochemical and biophysical mechanisms involved, and for her critical reading of the manuscript of this book. I am also indebted to Dr R. H. J. Brown for his advice and assistance in the technical aspects of experiments described here. Many of the results incorporated in this book were obtained while the author was supported by a special research grant from the Department of Scientific and Industrial Research, which is gratefully acknowledged.

J. E. H.

GIRTON COLLEGE
CAMBRIDGE
1964

## CHAPTER 1

# Introduction

THE study of biology is at a stage when it might be said that we know, on the whole, what living organisms do even if we do not know how they do it. The discovery, therefore, of a new facet of physiological and behavioural control is a rare event for the biologist. One of these rare events has been the discovery that plants and animals can measure time and can control their physiological processes and behaviour accordingly.

That animals and plants perform certain activities at fairly fixed times of day is not a new observation: no doubt the earliest hunters knew that certain animals were to be found drinking in the early dawn; they may even have noticed that mosquitoes bite more frequently at certain times of day. As early as 1729 De Mairan, an astronomer, described experiments in which plants were seen to maintain a diurnal rhythm of leaf movement although they were being kept in continuous darkness. Similar recordings were made by a number of authors in the nineteenth century, but such movements were regarded as being due either to an after-effect of exposure to the day:night cycle, or to an inherited tendency for movement.

Not until the 1930's was it fully realized that diurnal rhythms represent surprisingly basic periodicities, which are only temporarily, if at all, disturbed by abnormal light:dark cycles, and, most surprisingly, that such rhythms are not lost even after organisms have been kept in constant darkness for many generations. Although it was at this time that Wahl (1932) and Stein-Beling (1935) showed that bees possess a time-sense, diurnal rhythms were still regarded as quite separate phenomena. The term 'biological clock' did not come into common usage until Kramer, in 1952, showed that the sun navigation of birds is time-compensated. From this discovery dates the study of diurnal rhythms in terms of their relationship to an 'internal clock' system, and since then workers have laid particular emphasis on the functional prerequisites necessary for time-keeping, including reasonable independence from environmental variables, particularly temperature.

Pittendrigh and Bruce (1957a) have divided biological clocks into three categories:

(1) Continuously consulted clocks: those systems concerned with time-compensation when animals are navigating by reference to a celestial system.

(2) Interval timers: systems controlling discrete events such as onset of activity, or eclosion of an adult insect from the pupal case.

(3) Clocks controlling 'pure rhythms', that is rhythms in which the amplitude of a continuous process varies with time, for example the colour-change rhythm of crabs, or the rhythms of change of concentration of liver glycogen.

The controlling systems of all three categories, however, show the same characteristics; the distinction between them in fact lies in the type of 'indicator process' being observed. All three types of indicator process may be found in the one animal, and indeed even the distinction which is made between the different types of indicator process may depend only on how they are being measured.

Several models have been proposed for the 'clock' system (see Cold Spring Harbor Syposium, 1960), mainly based on Pittendrigh's (1959) model of two interacting self-sustaining oscillators. The use of models has stimulated certain lines of research and thereby fulfilled its purpose. Now that the properties of the clock system are better known, however, none of the proposed models any longer satisfies the requirements of the system; they are not, therefore, described in detail.

The history of any subject shows a waxing and waning in the interest aroused by different lines of investigation; a line may be pursued to a stage at which experiments, with only very slight modification, are constantly being repeated, then an advance in another line of investigation may cause total abandonment of the old line; however a new advance may distract attention from a previous line before it has been fully explored. The history of the study of diurnal rhythms, short though it is, follows just such a course. It is not the purpose of this book to follow each line in detail, but rather to survey in general terms the lines which have been followed, whether they have been rewarding or not, and to review them in the light of our most recent knowledge. It is only by considering the picture as a whole that we can see where fruitful lines of research have been abandoned too early, and where barriers against further advance have now been removed.

A choice has had to be made between indicating the large number of processes, within a vast range of organisms, which show diurnal rhythms (Harker, 1958a), and choosing a rather limited range of processes and organisms in order to review in detail the properties of rhythms. The latter alternative has been chosen, and as a result many well-known rhythms and experiments receive no mention, but it should be emphasized that the rhythms of the organisms quoted in the following chapters differ in no way from those of other organisms, and that the essential properties of the clock system appear to be very similar in all plants and animals.

## TERMINOLOGY

The use of the term 'diurnal rhythm' is open to criticism on the grounds of confusion with the term 'diurnal', meaning day-active as opposed to night-active. It has been used here simply because it is still in common usage and is known to most biologists as a term signifying a rhythm with a period of about 24 h. Since the discovery that animals in a constant environment hardly ever show a periodicity of *exactly* 24 h a new term, 'circadian rhythm' (from *circa* = about, *dies* = a day) (Halberg, 1959), has entered the literature. The use of this term is becoming more common, and it is here used interchangeably with the older term.

The terms used to describe various features of rhythms can most easily be defined by considering the rhythm as a quantity varying periodically with time, for example in the form of a sine wave (a gross oversimplification for any known diurnal rhythm) (fig. 1). The length of time taken to get from one point in the curve to the next identical point can be called the 'period' of the rhythm. Whatever form the wave takes, and whatever reference point is chosen, this measurement of period will be the same. Each particular point on the curve represents a particular 'phase' of the rhythm, and the maximum span of the quantity measured is the 'amplitude' of the rhythm.

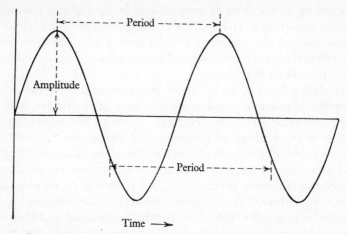

Fig. 1. Sine wave illustrating the terms used to describe rhythms.

The main difficulty in applying these concepts to diurnal rhythms may best be seen by considering the example of a particular rhythm, for instance that of the locomotor activity of a cockroach (fig. 2). The reference point for measuring the period of the rhythm might be taken as the time of onset of running activity, the time of the mid-point of running activity,

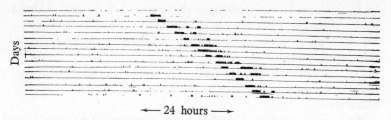

Fig. 2. Record of the running activity of the cockroach *Blaberus*. Recordings from successive days appear one below the other.

or the time of end of activity. The length of time for which running continues appears to be fairly irregular, and can in fact be shown to depend upon a number of factors unrelated to rhythmicity (e.g. age, amount of food present, the amount of activity on the previous day); the time of the mid-points of running activity also varies and is largely determined by non-rhythmical factors. The time of the end of activity is also irregular, but it can be seen that the onset of activity is quite regular

4

and would provide a suitable reference point for 'phase', and therefore for measuring the period of the rhythm.

It can be seen from this very simple example that the choice of the reference point for 'phase' may be critical, particularly when the effect of environmental variables on phase-timing is being considered. In many cases it is very difficult to separate those features of an apparently rhythmical process which are directly controlled by the internal clock from those which are not so controlled: if phase-timing is considered relative to processes which are only under the indirect control of the clock system most misleading conclusions can be drawn about the properties of the clock.

## CHAPTER 2

# The Environmental Control of Phase-timing

## LIGHT AS A PHASING FACTOR

MOST animals living in a natural environment are nocturnal, diurnal or crepuscular in their habits. This observation suggests that light is the most important of those environmental variables which affect the timing of the phases of a rhythm. In natural conditions, however, it is difficult to estimate which of the features of the day:night cycle determine the phase-setting of a rhythm.

The factors which might affect phase include (a) the increase in light intensity at dawn, perhaps bringing the intensity above some threshold level, (b) the decrease in light intensity at sunset, perhaps bringing the light intensity below some threshold value, (c) the intensity of the light during either, or both, day and night, and (d) the number of hours spent in darkness, or in light; furthermore it is possible that any combination of these factors together might be responsible for the establishment of phase-setting.

One way of testing whether the actual change from light to darkness, or darkness to light, affects the phase-setting of a rhythm is to subject an animal to a sudden reversal of the light:darkness cycle. Reversal of the cycle can only be achieved in one of two ways, (1) by keeping the light on for 24 h and then introducing a dark period at the time of day of the previous light period, (2) by maintaining the dark period for 24 h and then giving the first light period of the new conditions at the time of previous darkness. The way in which an animal responds to each of these two types of reversal will reveal to some extent which, if either, of the signals 'light on' or 'light off' acts as a phase-setter.

Chaffinches, which are day-active, have been subjected to this type of experiment (Aschoff, 1960). When the birds were exposed to two successive dark periods (i.e. given 24 h of darkness) only a very small peak of activity occurred at the normal time of day (fig. 3b), but another peak occurred when the light was turned on 12 h later than normal, despite the fact that this coincided with the chaffinches' subjective night. On the following days, during which the reversed light cycle was maintained, the active peak always occurred at the beginning of the light period. When reversal was achieved by imposing two consecutive light periods the birds became active at the time of their

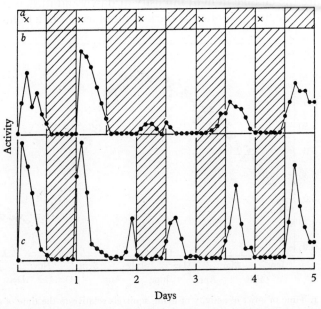

Fig. 3. Activity of chaffinches subjected to reversal of the day:night cycle of illumination through (b) two successive dark periods, (c) two successive light periods. (a) Time of previous light:darkness cycle, × represents the peaks of activity in such a cycle. (After Aschoff, 1960.)

first experience of a change from darkness to light, again despite the fact that this occurred during their subjective night (fig. 3c). The phase-setting of these animals is therefore at least partly determined by a change from darkness to light.

Similar experiments have been performed using a number of nocturnal animals including hamsters and cockroaches. The phase of the activity rhythm in these animals is found to be determined by the onset of darkness. So far then, it appears that, in those animals tested, phase is at least partly determined in diurnal animals by the time of onset of light, and in nocturnal animals by the time of onset of darkness.

## Abrupt and Slow Changes in Light Intensity

In the experiments just described the change in light intensity at the beginning and end of the dark period is large and abrupt, and may be thought of in terms of a light-on, light-off effect. In nature the change in light intensity at dawn and dusk is not an abrupt one, nor is it often as dark at night as it is in a light-tight

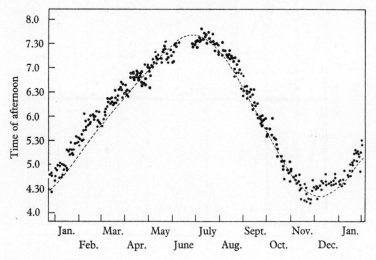

Fig. 4. Time of onset of activity of flying squirrels relative to the time of sunset. Dots represent activity onsets, dashed line represents time of sunset. (After DeCoursey, 1960.)

experimental chamber during the 'dark' period. Very few experiments have been performed using slowly changing light intensities, but observation of animals living in natural conditions indicates that a threshold value of light intensity is involved in the precise phase-setting of some rhythms.

A close correlation between the time of sunset and the departure from the den tree of a wild population of flying squirrels (*Glaucomys volans*) has been found by DeCoursey (1960). The same author kept flying squirrels in a large outdoor enclosure

8

where they could be more carefully studied, and again there appeared a close correlation between the time of onset of activity and a narrow range of light intensities at dusk (fig. 4).

In view of the above observations is is not surprising to find that the setting of the phases of a rhythm is by no means dependent upon a change from absolute darkness to light, or vice versa. Mori (1944) found that the phases of the rhythm of the extension and contraction of the sea pen *Cavernularia* could be set by alternating light intensities of 1230 lux and 130 lux. Fiddler

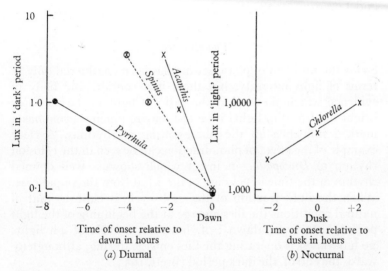

Fig. 5. (*a*) The effect of the intensity of the light during the 'dark' period on the timing of activity of three diurnal animals. (*b*) The effect on the timing of cell division (a nocturnal activity) of the light intensity in the 'light' period. (After Aschoff, 1960.)

crabs (*Uca*), subjected to a cycle of alternating bright and dim light, similarly show a correlation between the light cycle and the phases of the chromatophore rhythm. In the case of the crabs the setting of phase is related both to the relative increase in light intensity at the change from dim to bright light, and to the absolute intensity of the brighter light (Brown, Fingerman and Hines, 1954).

The intensity of the light during the 'dark' period also affects phase-setting in diurnal animals. Aschoff (1960) has

9

shown that the time of onset of activity of three bird species varies with the light intensity of the 'dark' period, even though the intensities tested were quite low, ranging from 0·1 to 10 lux. In all three species activity started earlier the brighter the light during the dark period. Another instance of an effect of the light intensity during the apparently inactive phase of a cycle is seen in the timing of cell division in the unicellular plant *Chlorella*: in this plant cell division occurs during the darker part of the environmental cycle but its exact timing is affected by the light intensity during the 'light' period (Pirson and Lorenzen, 1958) (fig. 5).

## Photofraction

So far the problem of phase-setting has been considered only in terms of light intensity, but there is a considerable body of evidence which shows that the ratio of light to darkness (or bright light to dim light) in a 24-h environmental cycle has a marked influence on the phase-setting of rhythms. A clear example of the effect of photofraction can be seen in the eclosion rhythm of *Drosophila*, an insect which shows a well defined rhythm in the time of eclosion of the adult from the pupal case. When the light fraction of the 24-h cycle is close to that of natural conditions the flies emerge at the beginning of the light period, that is at 'dawn'; if, however, a cycle of 4 h light: 20 h darkness is operating the flies emerge during, although towards the end of, the dark period (Brett, 1955).

Not all animals are as clearly affected by photofraction as is *Drosophila*; mice always show an activity peak at the beginning of both the dark and the light period, regardless of the ratio of light to darkness (provided neither falls below 4 h) (Aschoff and Meyer-Lohmann, 1955). This example, incidentally, introduces a case of a rhythm in which both the 'light on' and 'light off' signal is of importance in phase-setting.

The effect of photofraction is never a completely dominant one in relation to phase-setting; generally speaking in both plants and animals, whatever the ratio of light to darkness, the peak of the active phase remains close to the 'light on' signal in the case of diurnal organisms, and to the 'light off' signal in the case of nocturnal organisms.

There is a minimum of either the light, or the dark, fraction (but not both) below which the timing of the phases of the rhythms of most organisms lose their correlation with the environmental cycle; the organism may act either as though it were in constant conditions (see chapter 3), or it becomes arrhythmic. The phototactic rhythm of *Euglena*, for example, damps out when the light period extends beyond 20 h (Bruce and Pittendrigh, 1956), as does the oviposition rhythm of the mosquito *Aedes aegypti* (Haddow and Gillett, 1957, 1958). The fiddler crab ceases to respond to the phasing factor when there is only 1 h of darkness in the environmental cycle (Brown, Hines, Webb and Fingerman, 1950). The eclosion rhythm of *Drosophila* begins to break down if the light period is lengthened beyond 21 h, and the rhythm of sporulation of the fungus *Pilobolus* is affected in the same way (Brett, 1955; Übelmesser, 1954). On the other hand the rhythms of these latter two organisms are not affected by extension of the dark period and they remain extremely sensitive to very brief flashes of light; phase-setting can be determined in *Drosophila* by a light flash which lasts only 0·0005 s, and in *Pilobolus* by a flash lasting 0·005 s. In all the cases cited above it is notable that it is the extension of that part of the environmental cycle in which the activity normally occurs which causes breakdown of the rhythm. Several authors have drawn attention to this phenomenon and have considered it to be a significant character of rhythms. However instances are also known of rhythms which persist in both continuous light and continuous darkness, for example that of the locomotor activity of the cockroach *Leucophaea* (Roberts, 1959).

The action spectra for phase-setting are considered elsewhere (p. 56).

TEMPERATURE AS A PHASING FACTOR

Despite the relatively small effect of temperature on the period length of circadian rhythms (p. 23), cycling temperatures, within the range found in the normal habitat, may determine the phase-setting of some rhythms. Many authors have noted this as an extraordinary feature, but it is in keeping with the known action of light; light intensity has relatively little effect on period length, but changes in light intensity determine phase-setting.

11

The majority of organisms are, in respect of phase-setting, much less sensitive to cycling temperatures than to cycling light intensities, and temperature sensitivity may be revealed only when the light environment is held constant. On functional grounds this is not unexpected since temperature is a much more erratic environmental variable than is the change in light intensity over the 24 h.

Provided that the light conditions are constant a temperature cycle will determine phase-setting in animals as widely different as the protozan *Euglena* (Bruce and Pittendrigh, 1956), the beetle *Ptinus* (Bentley, Gunn and Ewer, 1941), the cockroaches *Leucophaea* and *Periplaneta* (Roberts, 1962; Cloudsley-Thompson, 1956), the rat (Browman, 1943; Calhoun, 1944), and the salamander (Pauli, 1926). The phase-setting of plant rhythms may also be determined by temperature, for instance those of sporulation of *Pilobolus* and *Oedogonium*, the latter being sensitive to a cycle with a range of only 2·5° C. On the other hand it has been shown convincingly that the phases of activity of the hamster and the flying squirrel are not affected by temperature cycles (Bruce, 1960; DeCoursey, 1960).

The phase-setting of the eclosion rhythm of *Drosophila* has also been reported to be sensitive to a temperature cycle (Pittendrigh, 1954), but the complication of the effect of temperature on developmental rates was not taken into account. However a rhythm may be *initiated* in unphased *Drosophila* cultures by a single 10° C rise in temperature lasting for 4 h, so that temperature almost certainly does affect phase-setting. The cockroach *Leucophaea* is also affected by only one experience of a change in temperature, a change of 5° C being sufficient to determine the time of running activity; running taking place at the time of maximum temperature (Roberts, 1959). This result is somewhat unexpected since the insect is nocturnal and in natural conditions would be active at a time of falling temperature.

The eclosion rhythm of the moth *Ephestia* is worth special mention, for it appears to be primarily controlled by temperature; a phase-setting which has been determined by a temperature cycle is not altered by a conflicting light:darkness cycle (Scott, 1936). The fruit fly *Dacus tryoni* also shows a sensitivity to temperature cycles, and an insensitivity to light cycles, when it is in the pupal stage (Bateman, 1955).

There is no doubt that the presence of food can act as a phasing factor for many rhythms. It is difficult to devise methods of testing this factor, and ideally experiments should include some test for hunger and digestion. If food is given to animals at only one time of day most animals will feed at that time, so in one sense the feeding activity is determined by the presence of food. Bees and ants undoubtedly show a rhythm of feeding in which the phases are determined by the time of day of the previous meal. A number of feeding times may occur within 24 h; the insects may visit feeding dishes at specific times and continue to do so even when they are not fed. Thus phase-setting in these insects can actually be brought about by feeding, and the rhythmicity is not simply related to a refractory period acting while digestion takes place. Bats also show a feeding rhythm in which phase is determined entirely by the time of the previous meal, provided that other environmental variables are held constant (Griffin and Welsh, 1937).

A number of investigations have been made of cyclical feeding of rats (Richter, 1927; Siegel and Stuckey, 1947; Verplanck and Hayes, 1953; Bare, 1959), and feeding is known to take place normally during the dark period. Rats will, however, change their running activity pattern if food is scarce and will feed during the day instead of, or as well as, during the night (Chitty and Southern, 1954). The change of pattern is not as complete as might be expected; Bare (1959) tested the effect of food deprivation, over varying periods of time, on the subsequent feeding rhythm, and found that the normal cyclical rate of intake was still maintained; that is the expected increase in intake following deprivation did not occur. If the time of feeding were entirely determined by the presence of food, then, after a period of deprivation the time at which food is first given would determine the time of the next feeding phase: this clearly does not happen.

When food is given to a cockroach only during the light period of a cycling environment the insect becomes active while the food is actually present, although the main peak of activity still occurs at the normal time, that is at the beginning of the dark period. If food is not given on any particular day there is, however, no sign of activity during the light period. Even in

13

continuous light the presence of food does not determine the phase-setting of the running activity, its presence, therefore, is not even a weak phase-setting factor (Harker, 1955).

### BAROMETRIC PRESSURE AS A PHASING FACTOR

Brown and his associates, in a long series of papers, have shown that there is a positive correlation between barometric pressure and the phases of a number of rhythms in the oyster, the crab *Uca*, the salamander, and many other organisms ranging from algae to vertebrates. The significance of these results is difficult to determine, and since they form a basis for a major controversy concerning the mechanisms of biological clocks they will be discussed separately in chapter 7.

### VARIATIONS IN PHASE-SETTING BETWEEN SPECIES

It is obvious that the phase-relationship with the light:dark cycle varies with different groups of organisms; different species of the same genus tend to have their own particular pattern of rhythm, and within the general pattern of the species individuals also vary. This suggests that the pattern of phase-relationship is controlled genetically, but only a few studies have been made of this aspect of circadian rhythms.

Two groups of female *Colias* butterflies, differing in one sex-linked gene, are known to fly at different times of day (Hovanitz quoted Kalmus, 1955), and several mutant strains of *Drosophila* have been shown to emerge from the pupal case at characteristically different times (Kalmus, 1955; Harker, 1964 *a*). The different activity patterns shown by the male and female of many species is also suggestive of a genetical basis. This sexual distinction is particularly clear in the phase-relations of insect emergence rhythms; the male generally appears to emerge up to 2 h earlier than the female in those species which have been studied (Caspers, 1953; Nielson and Haeger, 1954; Barnes, 1930). Sex-differences are also clear in the timing of the photo-tactic activity of flying insects (Williams, 1939).

### INTERACTION OF PHASING FACTORS

So far it has become obvious that light and darkness, temperature, the presence of food, and perhaps barometric pressure may each act as a factor determining the timing of the phases of circadian rhythms. In nature all these factors may be acting on an organism at the one time but in contrary ways. There is very little experimental evidence as to how organisms respond when

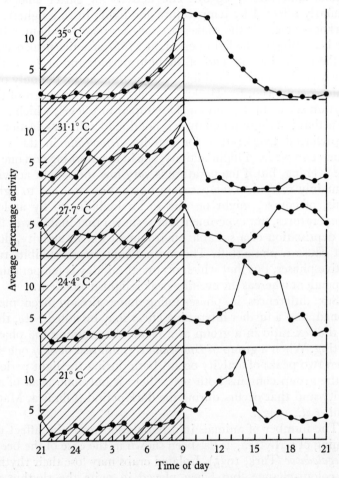

Fig. 6. The activity of a garter snake in alternating light and darkness at five different temperatures. (After Heckrotte, 1962.)

exposed to multiple 'directives', most of our knowledge being derived from observations made in natural conditions. Some of these rather random observations are cited below.

The interaction of temperature and light is known to affect

15

the ant *Messor semirufus*; the time of the peak of running activity occurs later in the day as temperature increases, occurring at noon in low temperatures, towards evening in slightly higher temperatures, and at night in still higher temperatures (Bodenheimer and Klein, 1930). The activity of garter snakes is similarly affected by temperature and light; in 12 h light:12 h darkness at 21° C the snakes are active about five hours after the beginning of the light period, at 27° C an active peak occurs at the beginning and end of the light period, and at 35° C the dawn peak is maintained for a longer time and is the only one which occurs (fig. 6) (Heckrotte, 1962). The expression of the rhythm seems to depend on the temperature level, which suggests that the rhythm in nature is not due to a direct selection of a preferred temperature for activity. *Drosophila* in the wild shows two peaks of flight activity, one in the morning and one in the evening, but if temperature is very low then flight activity is mainly related to temperature fluctuation and not to light. This, of course, might be due to an inability to fly at low temperatures, but experiments have shown that inability to fly, or deprivation of a suitable environment for an activity, does not cause a phase-shift. The moth *Plusia gamma* exhibits two active phases, but that which occurs at midday is temperature-dependent whereas the evening peak is light-dependent.

Sex differences in phase-timing have already been mentioned, and a further interesting effect may be noted here, that of the sex ratio in a group as a factor determining the phase-setting. When a group of *Drosophila* is composed of only one sex then two peaks of activity occur, both during the light period; if the group contains both sexes there is only one peak of activity and that occurs during the dark period (Ohsawa, Matutani *et al.*, 1942).

The number of animals in a population may also affect the timing of a rhythm, as has been shown in the case of the beetle *Megalodacne* (Park, 1935). Isolated crabs may lose their rhythm of colour-change, but when placed in pairs the rhythm reappears (Webb, Brown *et al.*, 1956). This observation suggests that the clock maintains its rhythm but the indicator process (the colour-change) drops below a measurable threshold value when the animal is isolated; stimulation, perhaps in the form of increased locomotor activity due to the presence of the other crab, causes the amplitude of the indicator process to increase to a level at which it is measurable. That the rhythm itself is in no way affected by the population effect is confirmed by observations made on groups of crabs in which individual members maintain rhythms which are out of phase with those of the rest of the group (Stephens, 1957 a). The time of emergence of *Drosophila* adults becomes later in the day as the population increases (Mori, 1954); this may be due to a change in the immediate environment, for instance increase in carbon dioxide concentration or decrease in oxygen. In fact many population effects may prove to be due to the effect of changes in the gas or food concentration in the environment.

The age of an animal affects the timing of the phases of its rhythms; this is particularly noticeable, not unexpectedly, in animals such as insects and amphibia, which move into a new environment at some stage of the life cycle. Adult insects, however, are also known to react differently to phasing factors as they age. The adult dragonfly *Anax imperator* shows an activity peak at dawn when it is newly emerged, and at midday when it is more than a week old: both age groups also fly at sunset (Corbet, 1957). Several species of Ephemeroptera are also active in the imago stage at different times of day according to age (Tjønneland, 1960); this is in contrast to the larval form which maintains throughout the larval period a phase-setting determined at a very early age (Harker, 1953).

The habitat of an animal may affect the phase-setting, in some cases to a considerable degree. The beetle *Feronia madida* is active only during the day in open ground, and only at night in woodland (Williams, 1959). Biting cycles of Culicidae are known to vary at different levels in a forest (Haddow, 1961), and as Corbet (1960) has pointed out the whole pattern of the physical factors of the environment can modify the phase-setting. The change in phase-timing with the above changes in environment may be partly due to the effect of hunger.

Finally in considering the interaction of phasing factors it is worth noting that very large numbers of species show diphasic or triphasic rhythms in the field, yet extremely few animals in the laboratory show other than a single phase of any activity. It seems likely that interaction of the factors which have now been shown to affect phase may be responsible for at least some of the multiple phases so often observed in nature. In the laboratory few, if any, experiments have been performed in which more than two environmental variables have been allowed to fluctuate, and this may account for the scarcity of experimental records concerning multiple phases. The fact that they *do* occur has been largely overlooked by authors dealing with possible mechanisms for timing, and particularly by those who have suggested model systems.

It can be seen from all the above evidence that the timing of the phases of circadian rhythms is influenced by a multiplicity of internal and external factors. It seems remarkable that an animal living in the fluctuating conditions of a natural environment should ever reveal a rhythm at all, but the generally overriding influence of light, together with the inherent limitations of phase-pattern within each species, must impose considerable control, for there is no doubt that very clearly defined rhythms, with remarkably small fluctuations in period length, are of common occurrence in nature.

## CHAPTER 3

# Free-running Rhythms

ONCE the phases of a circadian rhythm have been set by an environmental stimulus the rhythm will persist in favourable constant conditions, and may be maintained without any further stimulation for many months, or indeed throughout the lifetime of the animal. This may be so even when in the course of the lifetime the organism undergoes complete anatomical and

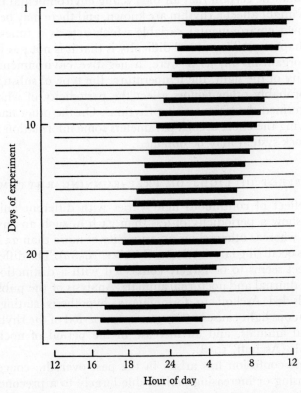

Fig. 7. Running activity of a flying squirrel in continuous darkness. Records from successive days appear beneath each other. (After DeCoursey, 1960.)

physiological reorganization such as that, for example, which occurs in the metamorphosis of a larval to an adult insect. In addition to the remarkable persistence of some rhythms there may also be extreme accuracy in the periodicity, for example the period of the running activity of the flying squirrel may be as accurate as ±2 min (which is an accuracy of about 0·1 per cent) (fig. 7).

It is this remarkable phenomenon which has led to the hypothesis that living organisms possess some type of 'internal clock'. Clearly one of the most important lines of investigation in the search for the 'internal clock' must be the attempt to define the properties of rhythms when organisms are in an environment which does not give time-cues. There are, how-ever, two basic difficulties in such an investigation; one is that we cannot be certain that all the cycling environmental factors which might affect a rhythm are known, and there may be some unknown environmental variable which acts as a time-signal (see chapter 7). The second difficulty is that it is not possible to keep organisms in, as it were, a negative environment: the intensity of the light, the temperature, the type of substratum, and probably other factors affect the periodicity of what has been termed the 'free-running rhythm'. On the other hand by observing the effect of such parameters some information about the clock system may be obtained.

EFFECT OF LIGHT ON FREE-RUNNING RHYTHMS

The effect of continuous light varies with different animals: some show a periodicity of less than 24 h in such an environ-ment, whereas others show a periodicity of more than 24 h, the phases occurring later each day (Harker, 1958 a). The difference in effect seems to be largely connected with a distinction be-tween diurnal and nocturnal animals. Analysis of the published records led Aschoff to formulate a hypothesis stating that continuous light causes a decrease in the period of the rhythm of diurnal animals, and an increase in the period of nocturnal animals (Aschoff, 1958).

Some confusion has arisen in the past over the concept of decreasing or increasing period, due largely to a preconceived idea that the natural, or free-running, period of an organism is exactly 24 h: thus any period greater than 24 h has been thought of as having lengthened, and any period of less than 24 h as having shortened. The recent practice (largely innovated by Pittendrigh and his associates) of leaving animals for many days in constant conditions in order to determine the free-running period has led to a better understanding of the whole problem. It is now fairly clear that even in continuous light the

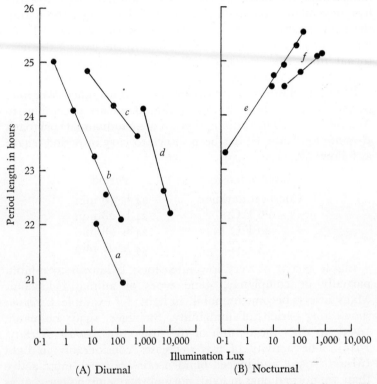

Fig. 8. The effect of light intensity on the period of rhythms of (A) diurnal (B) nocturnal organisms. (a) *Phaseolus*, (b) chaffinch, (c) lizard, (d) *Gonyaulax*, (e) *Mus*, (f) *Peromyscus*.

period of a diurnal animal may be greater than 24 h, transference to continuous darkness causing a further increase.

As might be expected, particularly in view of the effect of light intensity on phase-setting, the actual intensity of the light in a constant environment has a considerable effect on the length of the free-running period. Aschoff's hypothesis concerning the behaviour of diurnal and nocturnal animals in continuous light has been extended to include intensity effects, and this extended hypothesis has been widely called Aschoff's Rule. It states that with *increasing* intensity of illumination the period length of a diurnal animal decreases, while that of a nocturnal animal increases. The rule clearly holds in those cases cited by Aschoff (1960), which include the rhythms of activity of chaffinches and the lizard *Lacerta sicula* (Hoffman, 1960), of the leaf movement of the plant *Phaseolus* (Pfeffer, 1915), and that of the luminescence of the protozoan *Gonyaulax polyedra* (Hastings and Sweeney, 1958). It also holds in the case of the nocturnal mice *Mus musculus* and *Peromyscus leucopus* (Johnson, 1939; Aschoff, 1960) (fig. 8).

Rhythms of other species which have been studied over long periods of time show, however, that there are exceptions to Aschoff's Rule. The running rhythm of an individual cockroach *Byrsotria* for example, has been shown to vary in period length as follows (Roberts, 1959):

| Light intensity | Period |
|:---:|:---:|
| Constant darkness | 24 h 18 min |
| 0·6 F.C. | 24 h 28 min |
| 20 F.C. | 24 h 28 min |
| 25 F.C. | 24 h 21 min |

Light, except at very low intensities, is known to inhibit, partially or completely, some types of animal behaviour. Many insects become immobile in light; for example, *Carausius* shows long periods of immobility (Steiniger, 1936; Eidmann, 1955), as does the nymph of the dragonfly *Aeschna* (Serfaty, 1945). The activity of rats decreases considerably in light (Munn, 1950), and indeed blinded rats are much more active than are normal ones in light, normal rats being 99 per cent as active as blind ones in darkness, but only 48 per cent as active in light. In other instances darkness may inhibit some types of behaviour, and this is particularly noticeable in the emergence behaviour of arthropods. The chironomid *Pseudosmittia* will not emerge at all in darkness (Remmert, 1955), and in the red spider mite only 62 per cent of the normal emergence occurs (Hueck, 1951). Inhibition may continue to affect the animal even after the constant conditions have ceased, for instance *Convoluta* is considerably inhibited in its movements for some time after it has been removed from continuous darkness and placed in light (Bohn and Drzewina, 1928). The influence of other environmental characters operating at the same time as continuous light must also be taken into account. For example, the beetle *Ptinus* ceases to show an activity rhythm in continuous light, but the rhythm reappears if the humidity is suddenly increased (Ewer and Ewer, 1941).

Attention is drawn to this phenomenon as an extreme effect of constant conditions; it is reasonable to suppose that animals may in other cases be affected less profoundly, and even where obvious inhibition does not occur, the after-effects of constant light may continue for some time.

## EFFECT OF TEMPERATURE ON FREE-RUNNING RHYTHMS

Despite the fact that a temperature cycle may determine the phase-setting of circadian rhythms, the periodicity of such rhythms is only very slightly affected by temperatures lying within normal biological ranges. Indeed until fairly recently the period was thought to be completely temperature-independent, and much of the interest in rhythms has been due to this supposed temperature-independence.

Although it is now known that temperature-independence is not complete, the very small effect which temperature has on periodicity, compared with the effect which it has on most metabolic processes, is still remarkable. Most metabolic processes have a $Q_{10}$ of about 2 (that is the rate of the process doubles with a $10°$ C rise in temperature); the $Q_{10}$ of those circadian rhythms which have so far been investigated lie in the range of 0·8 to 1·3 (it being assumed that the rate of the process is inversely proportional to the period length—but see p. 26).

Brown and his associates have shown that the period of the rhythm of colour change in the crab *Uca* is independent of temperature over a $20°$ C range (Brown and Webb, 1948; Webb, Brown and Sandeen, 1954). This type of rhythm cannot be accurately measured in terms of minutes so that a very small effect might not be measurable, but the effect would have to be an extremely small one not to be noticeable over a long period of time.

Another case of apparently complete temperature-independence has been recorded in bees. Bees return to feeding stations at the same time of day with considerable accuracy; in view of the fluctuations in temperature in a natural environment they might be expected to be at least relatively temperature-insensitive. Controlled experiments by Wahl (1932) (fig. 9) and Kalmus (1934) confirm this expectation, and indeed Kalmus found temperature-independence throughout the range of 18° to 35° C. Again however the measurement of time-sense, or

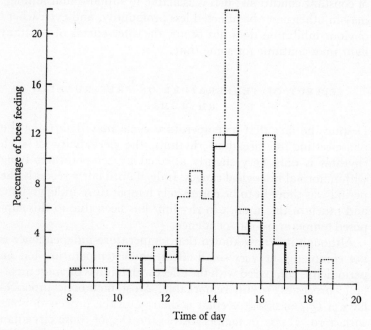

Fig. 9. The temperature insensitivity of the time-sense of bees trained to feed at a particular time of day. —— 31° C, - - - - 23° C. (After Wahl, 1932.)

periodicity, in these experiments is, of necessity, not accurate within small limits, and the very slight temperature effects recorded in other types of experiment might be obscured.

One of the clearest indications of a very slight temperature effect comes from Pittendrigh's observations on the eclosion rhythm of *Drosophila* when it is reared at three different temperatures (Pittendrigh, 1954) (fig. 10). Although great care must be taken in assessing the results of experiments on eclosion rhythms (see chapter 4) in this case there seems to be little doubt that the results do indicate a slight temperature effect.

Stocks raised at 16° C had an eclosion rhythm with a periodicity of about 24·5 h, while those raised at 26° C had a periodicity of 24 h, giving a $Q_{10}$ of 1·02.

Interesting results have come from the study of the effect of temperature on three different rhythms which run concurrently in the protozoan *Gonyaulax*. *Gonyaulax* shows a rhythm in the degree of luminescence which can be induced by a stimulus, a rhythm in the intensity of the spontaneous glow, and a rhythm

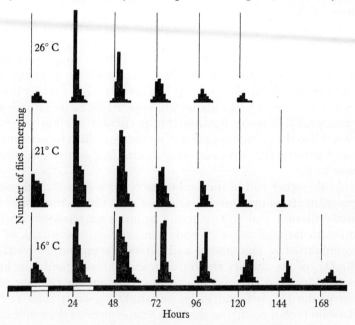

Fig. 10. Eclosion rhythm of *Drosophila* reared at three different temperatures. (After Pittendrigh, 1954.)

of cell division. The first two, induced luminescence and spontaneous glow, may be closely linked in mechanism, but the rhythm of cell division is unlikely to be directly linked to either. Yet all three rhythms appear to be very similarly affected by temperature, the periodicity of each having a $Q_{10}$ of between 0·85 and 0·9 (Sweeney and Hastings, 1958; Hastings and Sweeney, 1959). A $Q_{10}$ of less than 1 is, of course, fairly unusual since it means that the process is slowed down, rather than speeded up, by increasing temperature. The authors interpret this phenomenon as an evidence of overcompensation, and further suggest that, since all three rhythms show the same effect, the measured $Q_{10}$ is that of the actual controlling time-mechanism rather than being related to intermediary processes.

A similar slowing of the clock at higher temperatures occurs in another simple organism, the alga *Oedogonium* (Bühnemann, 1955 *a*, *b*). In this case the rhythm of sporulation has a period of 20 h at 17·5° C, 22 h at 25° C, and 23·2 h at 33° C, that is it shows a $Q_{10}$ of 0·8.

The periodicity of other unicellular organisms also shows a small $Q_{10}$, although larger than 1. The mating reaction of *Paramaecium* has a $Q_{10}$ very close to 1 (Ehret, 1959), and the rhythm of phototaxis of *Euglena*, measured at a range of temperatures from 16·7° C to 33° C, reveals a $Q_{10}$ varying from 1·01 to 1·1 (Bruce and Pittendrigh, 1956). This latter result is of considerable interest because this particular rhythm persists even when the culture is grown in a medium which induces such rapid growth that division occurs more frequently than once each 24 h.

At the other end of the evolutionary scale Rawson's elegant experiments on mammals have shown that the rhythms of both homeothermic and poikilothermic animals are relatively insensitive to large changes in body temperature. By means of a combination of anaesthesia and external temperature control the $Q_{10}$ of the activity rhythm of mice, hamsters and bats has been calculated by reference to their subsequent running activity: the $Q_{10}$ is found to be just slightly greater than 1. As Rawson (1960) points out the body temperature of these animals cannot be kept absolutely constant by the techniques employed, so that the calculations may not be as accurate as those which can be made with poikilothermic animals, but there seems to be no doubt that the $Q_{10}$ is of the order stated.

As has been stated earlier in this chapter the term $Q_{10}$ when used in connection with rhythms has always been calculated from measurements of the period of a rhythm. That is it has been assumed that the period is inversely proportional to the rate of the processes involved in the clock system. This assumption is difficult to justify while our knowledge is in its present state; it is, for example, possible that threshold concentrations and, or, decay values of substances determine the time at which the indicator process occurs, and we have no information as to

26

whether such values would be the same at all temperatures. This difficulty is further discussed in chapter 8.

The effect of very low temperatures presents another aspect of the clock system, and will be discussed in a later chapter (p. 51).

The findings described above, even after allowance for the way in which $Q_{10}$ has been calculated, are perhaps unexpected in physiological terms, for although it is possible to envisage a temperature-compensated clock mechanism it is remarkable that the metabolic and physiological processes, which must intervene between the clock and the indicator processes, do not have a more marked effect on the $Q_{10}$, especially in the case of complex behavioural rhythms. On the other hand the findings are not unexpected in functional terms, for if a biological clock is to serve any purpose as a time-keeper, then near temperature-independence is of primary importance: any time-keeper which doubled its rate with a 10° C rise in temperature (a relatively small temperature fluctuation in nature) would rapidly be at variance with solar time.

## EFFECT OF PREVIOUS ENVIRONMENT ON LENGTH OF FREE-RUNNING PERIOD

One of the least studied and understood, but nevertheless perhaps one of the most important effects of the environment is that of its long-term influence on the periodicity of rhythms. So far we have seen how the period of a rhythm is affected by the features of the prevailing environment, but such effects can be considerably modified by an after-effect from a preceding state.

Pittendrigh (1960) has demonstrated such an effect in an experiment in which he compared the period of the free-running rhythm in constant darkness of four sibling male hamsters after they had been kept first in a 23-h light:darkness cycle and then in a 25-h light:darkness cycle (fig. 11). It can be seen that the length of the period is considerably affected by the experience immediately prior to the free-running state. Experience of even a single 12-h light period by an animal otherwise kept in constant darkness similarly affects the periodicity (as well as causing a phase-shift).

A closer study has been made of the effect of light:darkness cycles on the free-running period of the cockroach *Blaberus*.

Individual cockroaches have been left in constant darkness for at least a fortnight, placed in a 12-h light:12-h darkness cycle for five days, and then returned to constant darkness. In every case the free-running period, on return to constant darkness after exposure to the light:darkness cycle, differed from its previous value. Furthermore the time of the first onset of light, in relation to the subjective night of the animal (= the estimated time of beginning of activity had the environment not

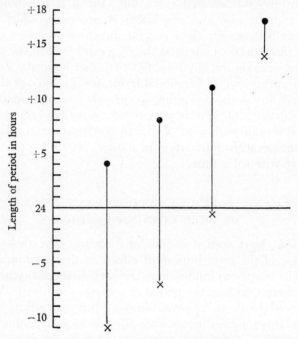

Fig. 11. The effect of exposure to 25-h days and 23-h days on the free-running period of four sibling male hamsters. ● = free-running period after 25-h day, × = free-running period after 23-h day. (After Pittendrigh, 1960.)

been altered), shows a close correlation with the ultimate percentage change in the free-running period (fig. 12). For instance if the light is turned on 4 h after the animal has begun to run about, the length of the free-running period when the cockroach is returned to darkness, is increased by 0·42 per cent; if the light is turned on 9 h after the active period the free-running period thereafter decreases by 0·63 per cent. The values for the percentage change in period are remarkably constant.

After environmental cycles of 18 h light:6 h darkness the percentage change in the free-running period is the same as that shown above. When animals are placed in constant light for five days and then returned to darkness the percentage change in period length is, however, very much greater, although the same relationship is shown between the time at which the light has been turned on, in relation to the time of subjective night, and the percentage change in period length.

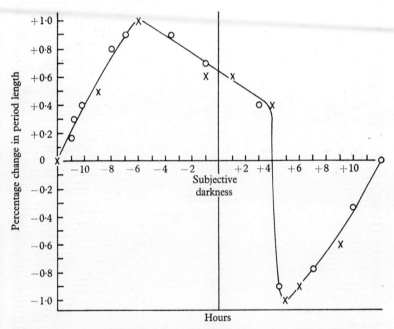

Fig. 12. The percentage change in the length of the 'constant darkness' free-running period induced by exposure to 12 h of light given at different times in the free-running cycle. O = *Blaberus*, × = *Leucophaea*.

The new period length is stable. Animals have been kept in constant darkness for at least a month after the experience of the cycling environmental conditions and throughout this time they maintained the new period length: two animals kept in darkness for three months showed no change in periodicity at the end of that time.

Calculations have been made, from the text figures given by Roberts (1959), of running activity of another cockroach,

*Leucophaea*, and a striking similarity between the after-effects in the two animals is apparent (fig. 12).

The results described above are of particular interest when they are considered in relation to previous concepts of the nature of circadian rhythms, for it has been tacitly assumed by most workers that the relative stability of the free-running period of any individual organism is an indication of the stability of the

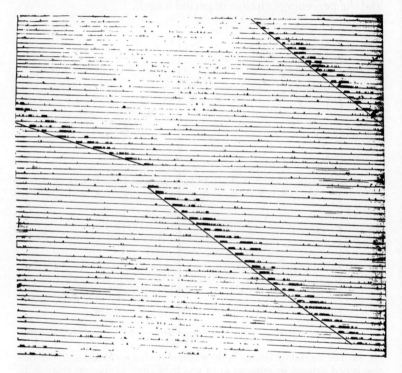

Fig. 13. Record of the running activity of the cockroach *Blaberus* in constant darkness showing a 'spontaneous' change in periodicity. Recordings from successive days appear beneath each other. A line has been drawn connecting the onsets of activity of successive days, the slope of the line shows the period of the rhythm.

clock system, and that variation in the periodicity of any one animal is due to the parameters of the constant conditions which are causing a slight decrease or increase in period relative to the natural period. This is now clearly seen not to be so, the so-called 'natural period' may be permanently altered by an exposure to light.

Change in the length of the free-running period is also known to occur occasionally while an animal is being kept in constant conditions; this type of variation has been called 'spontaneous change' in periodicity and has been recorded in both mammals and insects (DeCoursey, 1960; Rawson, 1960; Roberts, 1959) (fig. 13). It is possible that such spontaneous changes are produced by physiological changes within the animal, or that they are produced in the same way as are the sudden changes in behaviour pattern which occur as the seasons change or the animal ages. Menaker (in discussion of Rawson's 1960 paper) has found such a seasonal change in the free-running period of the temperature rhythm of bats; the period lengthening in winter and shortening in summer.

Very abrupt changes could possibly occur as a consequence of two or more rhythmical processes becoming dissociated. Dissociation of rhythms is known to occur under conditions of abnormal environmental cycles (p. 34), and constant conditions may act on an organism in the same way as an abnormal cycle. A dissociation of rhythms might itself produce a spontaneous change in the periodicity of an indicator process, or it is possible that when a particular phase-relationship between several dissociated rhythms occurs mutual entrainment may follow and be expressed as a change in the periodicity of the indicator process under observation.

Since some types of spontaneous change in period do not seem to occur with any regularity, or in relation to any particular set of conditions, it is possible that a combination of both the causes suggested above play a part in their production, and that it is only when an animal is in a particular physiological state that a spontaneous dissociation of rhythms occurs.

### LIMITS OF VARIATION IN PERIOD LENGTH

We have seen that the free-running period, whatever the parameters of the constant conditions, is always fairly close to 24 h, the extent of the variability being within the range of 23 h to 25 h in those rhythms described. It is clearly of importance that in the attempt to define the properties of the clock system it should be determined whether this range of period length is not only the greatest shown by free-running rhythms, but whether it in fact represents the limits of the period to which the clock can run. Since the phases of a rhythm are set by cycling environmental conditions it might be possible to change the period progressively by exposure to cycling environmental factors with a periodicity greater or less than the range shown by free-running rhythms.

Many studies have been made of the rhythms of organisms kept in environmental cycles having periods other than 24 h, and indeed the range of periodicities shown by such forced rhythms is wider than that found in free-running rhythms. On the other hand it is extremely difficult to interpret such evidence, for the reasons given below.

(1) The environmental cycle may be such that, although it has not a 24-h period, the disposition of its phases causes reinforcement of the organism's 24-h rhythm from time to time, causing major peaks at these times. Such an effect may give the impression that a change in the period of the rhythm has occurred. For instance a 24-h light:24-h darkness cycle may conflict with a persistent 24-h rhythm at every change from darkness to light, but it would also restimulate the rhythm at every change from light to darkness. Similarly the organism's rhythm might, every 48 h, come into phase with an environmental cycle of 8 h light:8 h darkness, or do so every 96 h with a cycle of 16 h light:16 h darkness. Since some physiological and behavioural states are actively inhibited by the presence of either light or darkness, rhythms involving such states may appear to follow the cycling environment exactly, although the periodicity of the clock system has in no way been altered from 24 h. Further difficulty is encountered in this sort of situation when the rhythm being measured is potentially a bimodal one: this difficulty has been practically ignored in the literature.

(2) The effect of a change in light intensity at, for the organism, an abnormal time of day is known to produce 'transients' in the periodicity (see chapter 4) and the effect of repeated induced transients is extremely difficult to interpret. In the literature however this aspect of the effect of environmental cycles has also been ignored.

(3) The organism may show a rhythmicity of some process, with the same period as the environmental cycle, but with its phase bearing a new relationship to the phase of the environmental cycle; this may suggest an abnormal response by the organism, but our knowledge is still so limited as to make it

unwise to be too definite in any interpretation of this type of result in relation to the clock system.

With these difficulties in mind some of the studies of variability of period length can be considered.

*Period Length in Environmental Cycles Longer or Shorter than 24 h*

The isopod *Ligia bandiniana* will take up a cycle of pigment dispersion correlated with an environmental cycle of 10 h light: 8 h darkness, but the pigment dispersion occurs during the dark phase and not during the light phase as it does in natural light cycles (Kleitman, 1940). There also appears to be a residual 24-h rhythm which persists despite the 18-h cycle.

A 9-h light:9-h darkness cycle is followed by the beetle *Anthei*, but not by *Blaps*; *Anthei*, however, reverts to an approximately 24-h cycle when placed in continuous darkness (Cloudsley-Thompson, 1956). The cockroach *Periplaneta* does not follow an 18-h cycle, but the tendency towards a bimodal rhythm is emphasized by the light:dark conditions of this cycle (Harker, 1958*a*).

Cycles of 8 h light:8 h darkness have been imposed on a number of animals. Some organisms will follow such a cycle while it is being maintained, but they revert to a 24-h periodicity when moved to constant conditions. *Euglena*, for example, follows a 16-h periodicity of phototactic sensitivity when in a 16-h environmental light cycle, but this does not persist in continous dim light (Pohl, 1948). The 16-h periodicity is notable in that the maximal phototactic sensitivity occurs during the dark period, and not during the light period as it does in a normal day–night cycle. The adult emergence rhythm of the moth *Ephestia* shows a 16-h periodicity if the pupae are kept in a 16-h temperature cycle (Scott, 1936), but there is evidence that a 24-h rhythm is still to some extent persistent. Some insects show considerable lability in period however: the emergence rhythm of the chironomid *Pseudosmittia* will conform to a cycle within the limits of 18 to 28 h (Remmert, 1955), and the cockroach *Leucophaea* will, in individual cases, follow cycles between the limits of 22 and 26 h (Roberts, 1959).

The white-footed mouse, *Peromyscus*, will not adapt to any light:dark cycle which deviates from 24 h by more than 3 h when the transitions between the two halves of the cycle are abrupt; it will however follow a 16-h environmental cycle if a slow twilight transition, in which the light intensity is halved every 3 min, is made (Kavanau, 1962). Furthermore, although the mice revert to a periodicity of close to 24 h when moved to constant darkness, the normal form of the rhythm is altered and two closely spaced major activity peaks are now evident.

The clearest method of showing just how far the period of a rhythm can be extended, and one which to some extent eliminates the possibility of overlooking the factors enumerated on p. 32, is that used by Tribukait (1954, 1956) in an investigation of the periodicity of the running activity of mice. The mice were originally kept in a 24-h light:dark cycle, and the period of this cycle was then gradually extended: the mice followed the lengthening environmental cycle until its period became greater than 28 h, after which the mice reverted to a shorter periodicity. As the period of the environmental cycle decreased steadily the mice followed the cycle until its period fell below 21 h. In contrast to these results the mice would not follow a 22-h or a 28-h cycle when this was imposed by a sudden switch from a 24-h cycle. Lizards will also respond to gradual lengthening of an environmental cycle, following one within the period range of 22 h to 26 h, although at the limits of the range individual variation is considerable (Liss and Frankel, 1958, quoted Bruce, 1960).

Attempts to modify the period of physiological rhythms in man have given results of particular interest. Taking advantage of the small variation which occurs in environmental conditions over the 24 h during the summer months in Spitzbergen, where perpetual daylight is experienced, Lewis and Lobban (1957 *a*, *b*) studied twelve human subjects living as two isolated communities. Seven of the subjects lived for seven weeks on the time-scale of a 21-h day, while the other five subjects pursued a 27-h day. Throughout the experimental period the rhythms of excretion and body temperature were measured. Whereas the body temperature almost immediately adapted to the imposed time-scale, the excretory rhythms were much slower to adjust, and at the end of the test period two of the subjects were still showing a 24-h rhythm while living on a 21-h time-scale. Although all those living on a 27-h time-scale showed some adaptation in water excretion the potassium excretion remained associated with a 24-h period. The main point of interest in these results is the dissociation of excretory rhythms; those of

34

water, sodium and chloride phase-shifted together, but that of potassium remained close to its previous phase-setting.

When all the above examples are considered it does appear that the periodicity of the clock may be entrained to as short an interval as 16 h or to as long an interval as 28 h, but it will be noticed that the rhythms which extend to the extremes of the range are those of either activity or emergence of adult insects: it is possible that in both instances the reactions recorded are immediate responses to an environment which overrides any signals from the internal clock system, and that the periodicity as measured may bear little relation to the period of the clock.

### Environmental Cycles which are Submultiples of 24 h

Environmental cycles of 12 h or less are not frequently followed by animals, but one of the changes from light to darkness may act as a phase-setter for a 24-h rhythm. Pittendrigh and Bruce (1957 b) have shown that *Euglena* shows a 24-h periodicity of phototactic sensitivity when a test light is lit for 15 min every 2 h, but the period of the rhythm alters to 23·3 h when the test light is on for 30 min every 2 h, despite the fact that the full cycle of the tests is still 24 h. If the test light is used for 15 min every $2\frac{1}{4}$ h the periodicity changes to either $22\frac{1}{2}$ h or $24\frac{3}{4}$ h. The authors maintain that these results are an example of frequency demultiplication, such as would arise if one oscillator were entrained by another, and if the period of the entraining oscillator (in this case the test light period) were exactly or nearly a whole or submultiple of that of the entrained oscillator (the endogenous rhythm of *Euglena*).

Pittendrigh and his associates have investigated this phenomenon further and have tested the effect of short light: dark cycles on the rhythms of the mouse *Peromyscus*, the hamster, and the cockroach *Leucophaea*. In all cases periodicity of close to 24 h occurs under such environmental conditions.

The term frequency demultiplication is now common in the literature pertaining to circadian rhythms, and although to some extent it serves a useful purpose in providing a short term for the phenomenon involving the production (persistence?) of a circadian rhythm by an environmental cycle in which the period length is a submultiple of the period of the induced rhythm, it serves, on the other hand, as an excellent example of the way in which the use of a technical term can obscure and complicate our understanding of a fairly simple phenomenon. All the organisms which show the so-called frequency demultiplication effect have also been shown to be practically unresponsive to short periods of environmental stimulation when these occur during the subjective day: for example the flying squirrel is not responsive to short light periods imposed during its subjective day, even when the animal has been for a considerable period in continuous darkness. The lack of response, or sensitivity, of these organisms to environmental changes over some part of the 24 h, and the limitation in the degree of phase-shift which can be induced by environmental stimulation in the other part of the 24-h cycle (see p. 44) suggests strongly that short period light:dark cycles only affect the animal when they occur at about the time when the animal would in any case become active. Therefore the use of the term 'frequency demultiplication', with its implication of entrainment of oscillating systems, is an unnecessary and misleading complication.

CHAPTER 4

# Phase Perturbation

## PHASE-SHIFT INDUCED BY NORMAL
## ENVIRONMENTAL CONDITIONS

WHEN the phases of a circadian rhythm are shifted by changes in the timing of the phases of the environmental cycle the period of the rhythm must, at least temporarily, be affected. Observations of the effect, on an established rhythm, of sudden changes in the environmental cycle should therefore give some further information about the possible lability of periodicity of the clock; moreover such observations should give a certain amount of information about either the responsiveness of the

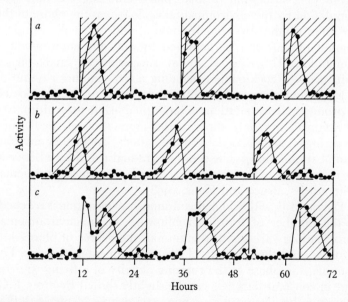

Fig. 14. The effect on the timing of the activity rhythm of the cockroach *Periplaneta* of (b) starting the dark period earlier than usual, (c) starting the dark period later than usual; (a) control in normal conditions.

clock throughout the 24-h cycle, or the condition at various times of day of those intermediary processes which lie between the 'clock' and the indicator process.

Figure 14 shows the effect of a change in the timing of the light:dark cycle on the running activity of the cockroach *Periplaneta* when the beginning of darkness comes later than usual, and when it comes earlier than usual. Information about both the periodicity of the rhythm, and about the condition of the clock, can be gained from even this simple experiment, for it can be seen that (1) the time of the running activity moves (phase-shifts) so as to lie in its usual relationship with the beginning of darkness, but it does so through a series of phase-shifts on successive days; thus the period between successive phases of activity never falls outside the range already shown to be the limit of variability in period, (2) the direction of the phase-shift differs in the two cases: when darkness comes earlier than usual the active phase occurs earlier each day, when darkness comes later than usual the active phase occurs later each day. The way in which a large phase-shift is achieved by means of a succession of small phase-shifts suggests that each stimulation by the environmental cycle moves the phase of the circadian rhythm by a small amount, and that only a limited amount of shift can be achieved by any one environmental stimulation. However a similar succession of small phase-shifts occur if cockroaches pursuing a free-running rhythm in constant environmental conditions are given but one unrepeated environmental stimulation: a large phase-shift may ultimately be attained, but only through the summation of small shifts occurring on successive days (despite the lack of further environmental stimulation). The series of gradual transitions between the old and the new phase-setting have been termed 'transients' (Pittendrigh, Bruce and Kaus, 1958).

The rapidity with which the ultimate phase-setting is reached, or the number of transients following on an environmental perturbation, varies with the animal, the degree of phase-shift involved, and the intensity of the environmental stimulus. For example, the phase of the running activity of a flying squirrel kept in constant darkness can be slightly shifted by a 0·5 F.C. light stimulus lasting for 10 min; the degree of phase-shift is increased by increasing the length of the light period, the maximum possible phase-shift being induced by a 50-min light period (DeCoursey, 1960). In these circumstances the transient effects, if any, are extremely small. On the other hand a repeated 12-h light:12-h darkness cycle, when the light period is initiated during the subjective night of the previously free-running rhythm, causes the phase of running activity to phase-shift

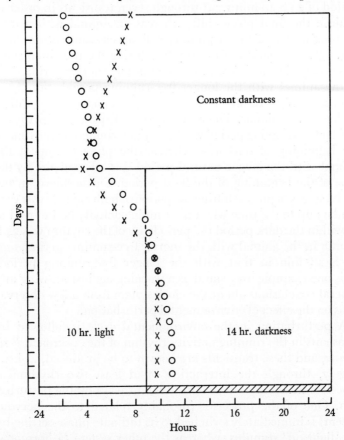

Fig. 15. Diagrammatic representation of the effect of a light: darkness cycle on the running activity of a flying squirrel with (a) a free-running period of 24 h 15 min (○), (b) a free-running period of 23 h 40 min (×). (After DeCoursey, 1960.)

through 7 h until it again lies in its normal relationship with darkness. In this case as many as fourteen transients have been reported.

Individual variation in the response to environmental perturbation can also be seen in the results of DeCoursey's experiments with flying squirrels. In all cases when a light cycle of 10 h light:14 h darkness was given to animals previously kept in continuous darkness, the light period being initiated at the time of subjective night, the running activity phase-shifted to lie within the dark period, but an animal with a free-running period of 23 h 40 min went through seven transients in order to achieve the final phase-shift, whereas an animal with a free-running period of 24 h 15 min took fourteen transients to do so (fig. 15). The period of the rhythm during the transition stage was greater in the animal with the initially shorter period than in the animal with the longer free-running period, but closer inspection of the results shows that the final phase-setting differs in these two animals. The phase-setting of the animal with the shorter free-running period was finally determined by the time of the beginning of darkness, whereas the phase-setting of the animal with the longer natural period was affected also by the time of the beginning of the light period. Furthermore, in the initial stages of phase-shifting the period of both animals was very similar; up to the time when the running activity had moved to lie within the dark period the periodicity of the rhythm was 24 h 55 min in the animal with the shorter free-running period, and 24 h 48 min in that with the longer free-running period. This one example may stand as a warning against drawing any general conclusions about the clock system from a few observations on the effect of environmental perturbation.

A perturbation in the environmental cycle is followed by transients in the running activity rhythm of the cockroach *Periplaneta*, and these transients are known to be produced, at least largely, through the interaction of at least two rhythmical processes which are affected in different ways by the perturbation. One of the processes (probably related to the nervous system) is immediately switched to its ultimate phase-setting by the 'light off' stimulus, whereas the other system (a hormonal one), although affected by the first system, phase-shifts only gradually but it continues to do so until both processes come to lie in phase with each other. It can be seen that the interaction of these two systems allows for the development of transient responses in the rhythm of the indicator processes, regardless of whether the environmental perturbation is continued or occurs only once. The details of the two systems are described in chapter 5.

In some higher animals it has been possible to effect differential phase-shift in several processes, for example the excretion of potassium and water in man (see page 34), and this effect, together with the differential effects of light intensity and photofraction on phase-setting, strongly suggests that a number of rhythmical processes may be controlled by different centres, and that, as in the case of the cockroach, observed transients of one indicator process may be the result of mutual interaction between several controlling centres, or the processes which they regulate, the final phase-setting being achieved only when they become stabilized relative to each other.

In view of such possibilities it is of particular interest to examine the effect of environmental perturbation on the rhythms of unicellular organisms. Here the possible number of separate control centres for the various rhythms is at least decreased, and in *Gonyaulax*, the most intensively studied of such organisms, at least two characteristically rhythmic functions, those of luminescence and cell division, cannot be separately phase-shifted (Hastings and Sweeney, 1958). Unfortunately the study of the effect of perturbations on the production of transients has not so far produced any very clear cut result. In *Gonyaulax* the initial phase-shift is said to be a stable one, with no transients involved (Hastings and Sweeney, 1958), although comparable effects in *Euglena* were described as transients by Bruce and Pittendrigh (1956). In the case of the phototactic rhythm of *Euglena* what is called a transient approach to a new phase position appears to be largely due to a very small change in amplitude of the phototactic response. The difficulty in interpretation arises because the amplitude of the response, rather than the time of the response, is the critical measurement, and amplitude is a notoriously difficult parameter of rhythms to interpret.

### DIRECTION OF PHASE-SHIFT AFTER ENVIRONMENTAL PERTURBATION

Attention has already been drawn to the fact that the effect of a light stimulus on the phase of the free-running activity of a cockroach depends on whether it comes before or after the time of the cockroach activity, that is before or after the beginning of the cockroach's subjective 'night'. Such observations may yield a considerable amount of information about the clock system.

The effect of exposure to short periods of light (3 h) given at different times in the cycle of the free-running rhythm of luminescence of *Gonyaulax* can be seen in fig. 16. The results suggest that there is a period during the subjective 'day' when the organism is unresponsive to light; at other times the phase is

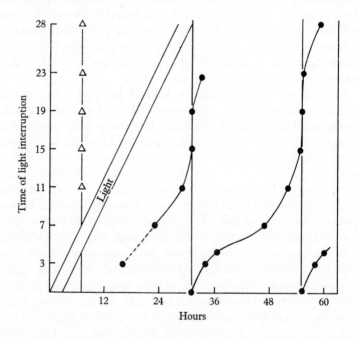

Fig. 16. The effect on the free-running rhythm of luminescence of *Gonyaulax* of a 3-h light period given at different times of day. Cells had been kept in 12 h light: 12 h darkness prior to the beginning of the experiment. Zero time on the graph represents the end of the final light period. Circles represent the times of maximum luminescence. Triangles and vertical lines show time of normal luminescence if no light interruption is given. (After Hastings and Sweeney, 1958.)

shifted, but the new phase may be achieved either by a temporary shortening of the period of the rhythm (advance reset), or by a temporary lengthening of the rhythm (delay reset) (Hastings and Sweeney, 1958). In Hastings and Sweeney's figure (1958) of an experiment in which the light perturbation occurs 3 h after the beginning of subjective night (line 3) the first peak appears 28 h later than usual. In a similar experiment described elsewhere (Hastings and Sweeney, 1959) a small peak is clearly seen 24 h earlier than this: this peak is represented in fig. 16 by the dotted line.

The type of effect seen here can be considered in two ways. It can be seen that phase-shift is achieved by an advance reset when the light signal occurs between the time of maximum luminescence and the beginning of subjective 'day', and that phase-shift is achieved by a delay reset when the light signal has been given during the first part of the subjective night. If the phase-shift was not achieved in this way, but say in the reverse manner (delay resets after a light signal during the late night period, and advance resets after light during the early night period), then the transition between the old and the new phase would involve either a great many transients or a periodicity widely divergent from 24 h: from the previously described evidence the latter effect does not seem likely.

Another way of regarding these results is to consider them in respect of possible ways in which the clock system might work. It may be that some threshold value is reached every 24 h, or that some of the intermediate steps between the clock and the indicator process involve metabolic stages in which substances build up to a threshold level. It is then possible that if the light stimulus prevents expression of the indicator process the process would occur as soon as the inhibition ceased (i.e. a delay response). If, however, the light comes much earlier than usual then the stimulus for luminescence, which appears to be the 'off' signal, would come too soon for the expression of the process to be possible, but the process would occur as soon as the threshold values had been reached, and the result would be an advance response.

Very similar responses to light perturbations are shown by many different types of animal including *Euglena*, *Paramaecium*, flying squirrels, hamsters and cockroaches. Pittendrigh has pointed out that all these organisms show either advance or delay shifts when a light signal is given at appropriate times of the cycle, and in all species the switch from the delay to the advance response comes during the subjective night. Response curves for some of these animals are given in fig. 17. If it is recognized that the period of the rhythms being measured is labile only within a small range of 24 h then the similarity in the response curve of all animals is not surprising. Continued phase-shift by a delaying response as the light signal came closer and closer to the normal time of subjective day would force the periodicity into larger and larger deviations from 24 h. If the switch from an advance to a delay response comes at the time of maximum possible deviation from the 24-h period then we would expect the results of experiments in which animals were kept in continuous darkness and then moved to continuous light to give a

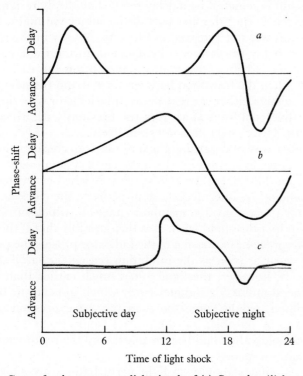

Fig. 17. Curves for the response to light signals of (*a*) *Gonyaulax*, (*b*) hamster, (*c*) flying squirrel. Response is measured in terms of change of phase of the rhythm when light signals are given at different times of 'subjective day'. Onset of activity prior to light signal has been taken as occurring at 12 on the abscissa.

response curve exactly the reverse in form to that of the light-response curve. This does occur, at least in the case of the cockroach *Blaberus* (fig. 18).

Really close comparison between the responses of species cannot be made (in any case the individual variation may be large) because the observations so far recorded have involved a wide variety of light intensity and length of light period, both of

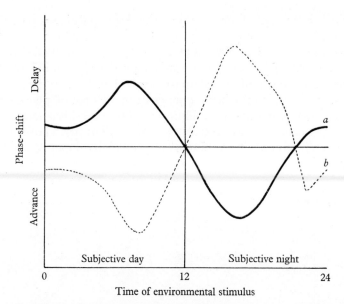

Fig. 18. Curves for the response of the cockroach *Blaberus* to (*a*) short light signals given at various times of 'subjective day' ——, (*b*) changes from constant darkness to constant light at various times of 'subjective day' – – – – .

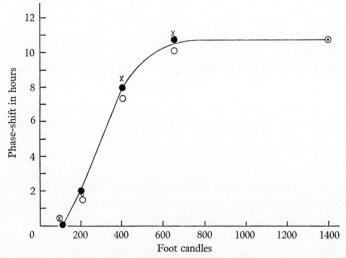

Fig. 19. The effect of light intensity on the degree of phase-shift in the rhythm of luminescence of *Gonyaulax* induced by a $2\frac{1}{2}$-h light signal given 6 h after the beginning of subjective night. ○ = first maximum, ● = second maximum, × = third maximum. (After Hastings and Sweeney, 1958.)

which have been shown to affect the degree of phase-shift. For example, when a light perturbation is introduced 6 h after the beginning of subjective night the rhythm of luminescence of *Gonyaulax* is phase-shifted by an amount proportional to the duration of the light period up to a limit of $2\frac{1}{2}$ h, when the phase-shift is $11\frac{3}{4}$ h; no further increases in the degree of phase-shift results from prolonging the light period beyond this length. The effect of different light intensities when a light period of $2\frac{1}{2}$ h is given 6 h after the beginning of subjective night can be seen in fig. 19.

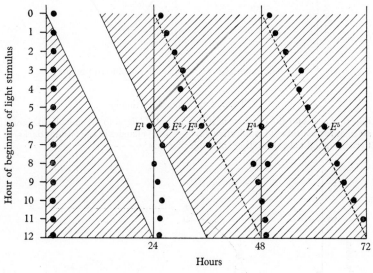

Fig. 20. The effect on the rhythm of eclosion of *Drosophila* of exposure to a 12-h light period beginning at different subjective times for each culture (each horizontal line represents a separate culture). The circles represent major peaks of emergence. (After Pittendrigh, 1959.)

## Transients in the Eclosion Rhythm of Drosophila

The most frequently quoted case of the production of transients by an environmental perturbation is that occurring in the eclosion rhythm of *Drosophila*: the Pittendrigh oscillator model has been largely based on the conclusions drawn from the results of these experiments.

In one set of experiments Pittendrigh (1958) took thirteen separate cultures of *Drosophila* pupae which had previously been kept in 12 h light: 12 h darkness; after a period of 24 h in constant darkness he exposed each of these cultures to a 12-h light period, the light period beginning an hour later in each successive culture: thereafter the cultures were left in continuous darkness. Fig. 20 shows the median points of eclosion in individual cultures throughout the treatment. Pittendrigh has affirmed that the results are comparable to those obtained for a repetitive rhythm, and that the results are in no way obscured by the fact that the system is a population (Cold Spring Harbor Symposium, 1960, p. 183).

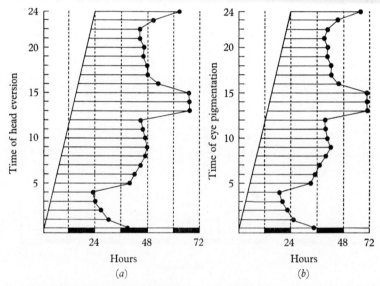

Fig. 21. The time taken for development of *Drosophila* pupae from the stage of (a) head eversion to yellow eye coloration (b) from yellow eye coloration to costal vein pigmentation, relative to the time of day at which the initial stage is reached.

Recent studies (Harker, 1964a) however underline the dangers of such an approach. A study has been made of the developmental process of *Drosophila* throughout the pupal stage, and it has been found that the rate of development is affected by light:dark signals, and by rhythms which are free-running. The time taken for the pupa to develop from any one of four stages to the next stage depends on the subjective time (i.e. in relation to phase-setting) at which the pupa enters the stage being observed. For example, as can be seen in fig. 21a, when pupae are kept in 12 h light:12 h darkness the time taken for a pupa to develop from the stage at which head eversion takes place to the stage at which the yellow eye coloration can be seen differs considerably when head eversion occurs at different times of day: the minimum period for development is 19 h, and this occurs when the head everts 4 h after the beginning of the light period: the maximum period is 54 h, and this occurs when the head everts one hour after the beginning of the dark period. A similar relationship between the light:dark cycle and the rate of development from the 'yellow eye' stage to the 'costal vein pigmentation' stage can be seen in fig. 21 b. Again the

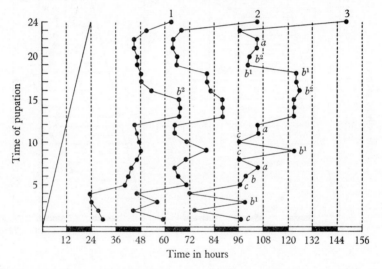

Fig. 22. The time required for development from the beginning of head eversion to eclosion for *Drosophila* pupae reaching the first stage at different times of day. Vertical lines 1, 2, 3 represent the three stages head eversion to yellow eye, yellow eye to wing pigmentation, wing pigmentation to eclosion. For symbols $a$, $b^1$, $b^2$ etc. see text.

minimum period involved (15 h) occurs when eye pigmentation begins 4 h after the beginning of the light period, and the maximum period (44 h) when eye pigmentation begins an hour after the beginning of darkness. Similar results are obtained even when the pupae are kept in continuous darkness throughout the experiment, provided that at some time during the larval stage they have been in a 12-h light:12-h darkness cycle.

Because pupation occurs at random throughout the entire

48

24 h, and the rate of development to the head eversion stage is fairly constant, pupae enter that stage at all times of day. As a result of differential growth rates from that stage on, pupae may reach one stage rapidly but develop slowly to the next stage; the result of this on the times of eclosion of the culture as a whole is to produce a circadian rhythm with a large peak of eclosion soon after dawn (fig. 22). The eclosion rhythm is therefore a population effect rather than the reflection of phasing of individuals to a dawn eclosion; individuals can, and do, undergo

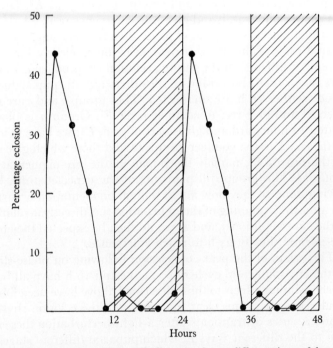

Fig. 23. The percentage of *Drosophila* flies emerging at different times of day calculated by summing the daily emergence shown in fig. 22.

eclosion at nearly all hours of the day, but the majority emerge at dawn because of the summation effect of circadian rhythms at earlier stages (fig. 23).

If a light perturbation of 12 h is given to a group of pupae kept in constant darkness, but phased to a 12-h light:12-h darkness cycle earlier in their life cycle, a very complex pattern of phase-shift results and this can only be resolved by following

4           49

the development of individual pupae. The whole pattern is too complex to be dealt with in detail here, but some brief generalizations can be made. If the light signal occurs 12 h out of phase with subjective time, that is at the beginning of the subjective night (for example at hour 84 or 108 on fig. 22), then any pupa which at that time is within 15 h of its expected eclosion time is unaffected and emerges at the expected time. These pupae are termed $b^1$ and $c$ in fig. 22, and in fig. 20 represent the eclosion peaks $E^2$ and $E^1$. The group $b^2$ and some of $b^1$ show a delay of 8 h and emerge to give peak $E^3$. Group $a$ is phase-shifted 15 h and gives the emergence peak $E^4$. If the light signal occurs 14 h out of phase then the resulting developmental rates of the pupae reveal even more clearly how apparent 'transients' in the ensuing eclosion are produced. In this case, as in that above (and in all others tested), pupae due to eclode within 15 h are unaffected. Thus group $b$ and $c$ are unaffected and give rise to eclosion peak $E^1$. Group $a$ is phase-shifted 12 h forward and gives part of peak $E^2$: we have to look further back to the younger groups to find those which give rise to the rest of $E^2$, namely the pupae in the pre-pigmentation group $b^2$. This group still pigments at the expected time, but having done so they now act as though pigmentation occurred 3 h after the beginning of the light period (i.e. directly in relation to the new conditions), and as would then be expected they take the normal time of 24 h to reach emergence.

The origin of the peaks of eclosion following on phase-shifts in the environmental cycle of 6, 12, 14 and 16 h have all been traced; similar effects to those described above have been found in all cases. It is thus clear that although the eclosion rhythm shows a series of transients after a light perturbation these are due to the different ways in which pupae at different stages of development are affected, the population being made up of mixed age groups; they are not due to transient effects in individual rhythms of emergence.

## PHASE-SHIFT INDUCED BY ABNORMAL ENVIRONMENTAL CONDITIONS

Although the period of circadian rhythms cannot be altered, beyond certain narrow limits, by the environmental factors which normally act as phase-setters extreme environmental conditions such as temperatures near freezing point, or light of very high intensity, may affect rhythms more drastically. Altering the internal environment of the organism, when extreme measures are taken, may also affect its circadian rhythm to some degree. Such affects are generally apparent only if the imposed conditions are such that the rhythmical process cannot function while the conditions are maintained, and the effect on the rhythm can therefore only be measured in terms of phase-shifts or alterations in period length when the environmental conditions are returned to normal.

## The Effect of Low Temperature

The effects of low temperature on rhythmical processes can be divided roughly into five categories:

(1) After return to normal temperature the phase of the rhythm may appear to have shifted, relative to solar time, by the same interval as that spent by the organism at the low temperature. That is the animal behaves as though the internal clock had stopped during the low temperature conditions and had been restarted on return to higher temperature.

This type of phase-shift has been observed in the rhythm of colour change of the crab *Uca* after it has been kept at 0° C (Brown and Webb, 1948).

(2) The phases of the rhythm after return to normal conditions may be shifted by a set number of hours, not equal to the time-interval spent at the lower temperature. This type of behaviour has also been observed in *Uca*; it occurs when the temperature is lowered to only 9·5° C. In this particular case increase in the time spent at low temperature beyond 12 h has little further effect on the degree of phase-shift (Stephens, 1957*a*, *b*); for example, after a temperature drop lasting 12 h the phase delay is 3 h 20 min, after a temperature drop lasting 84 h the phase delay is 3 h 29 min.

(3) The degree of phase-shift may be directly related to a fixed time-interval measured from the time of return of the organism to normal temperature. This type of effect has been described in *Gonyaulax*; in this protozoan the rhythm of luminescence ceases at 11·5° C, it is regained 30 h after *Gonyaulax* is transferred back to a higher temperature (Hastings and Sweeney, 1957*a*). It is also notable that in this organism the rhythm is lost altogether after the temperature has been lowered to 6° C and

then raised, despite the fact that the organism regains its ability to luminesce at the higher temperature.

(4) The degree of phase-shift may be directly related to the stage of the cycle at which the low temperature is introduced. This type of effect has been described in *Periplaneta* (Bunning, 1958, 1959). Cooling for a short period, beginning near the time of the active phase, causes a large shift in the final phase-setting, whereas cooling for a short period during the inactive phase causes very little shift in the phase position (fig. 24). Similarly low temperature causes a maximum phase-shift in the

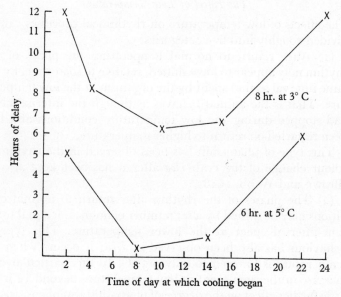

Fig. 24. The effect of a period at low temperature, starting at different times of day, on the time of activity of *Periplaneta* when it is subsequently returned to normal temperature. (After Bunning, 1959.)

activity rhythm of lizards when it occurs at the time of maximum activity (Hoffman, quoted Sweeney and Hastings, 1960). In the crab *Uca* there is also a maximum shift in the melanophore rhythm when the cold period is initiated at the time of the peak of the rhythm (Stephens, 1957*b*).

(5) There may be only an extremely small shift, or no shift at all, in the phase-timing after a period spent at low temperature. For example the effect of low temperatures on mammals which

hibernate is remarkably small: Rawson has shown that the reduction of the body temperature down to 5° C causes a delay in the running phase only of the order of a few minutes (Rawson, 1960).

It seems likely that the different types of reaction listed above do not reflect five different types of clock system but are due to the temperature-dependence of each of a multiplicity of reactions which together result in an observed behavioural rhythm. Evidence has already been given in support of the hypothesis that there may be several centres controlling rhythmical processes and that these centres, or the processes controlled by them, may interact with each other. It is possible that different centres may have different threshold values for low temperature resistance, and that when the temperature is lowered below all threshold values the animal shows a phase-shift equal to the time spent at the lower temperature. If, however, the temperature is lowered to a degree which is below the threshold of only some, or only one, process then those rhythmical processes which are unaffected by the temperature drop may, by mutual entrainment, influence the degree of phase-shift of those which have been affected by the temperature drop. It is also possible that intermediate processes between the clock and the indicator processes are differentially affected in different animals; such processes may not be rhythmical in themselves and any effect which they may have on the timing of the indicator process may give misleading information if considered in terms of the mechanism of the clock system.

## The Effect of Anaesthetics

Surprisingly little work has been done on the effect of anaesthetics on rhythms, despite the fact that the effect of anaesthetics to some extent mimics that of low temperature, and that by using anaesthetics which affect different systems some indication might be gained of the pathways through which the clock system acts.

There have been claims that carbon dioxide anaesthesia causes a phase-shift of the same type as that caused by very low temperature, the phase being shifted by the same number of hours as has been spent under anaesthesia. Nitrogen has also been reported to have a similar effect in some cases. *Drosophila* cultures, after being kept in nitrogen for 15 h, show a phase-shift of 15 h immediately after the treatment, but the stable phase-shift is only one of 10 h (Pittendrigh, 1954). It seems very likely that the apparent transient effect in this case is due to a differential effect of nitrogen on two different stages in the development of the *Drosophila* pupa.

Modification of the activity rhythm of *Periplaneta* has been induced by treatment with carbon dioxide and nitrogen (Ralph, 1959). The effect on the rhythm of the exposure to nitrogen depends on the stage of the cycle at which it is given. If the animal is exposed to nitrogen 2 to 4 h before the expected peak of activity there is no phase-shift in the subsequent rhythm, but if it is given at 10 or 22 h before the activity peak then there is a shift in the subsequent phase-setting, the time of activity being delayed. The closer the time of nitrogen treatment to the last active peak the greater the subsequent phase-shift. This is the opposite effect to that of low temperature, in that case it will be remembered that maximum phase-shift occurred when the time of low temperature came closest to the time of the activity peak. It seems likely that nitrogen anaesthesia affects some particular process in the rhythmical cycle which occurs after the time of activity, it may for example act on the subsidiary endocrine cycle which follows the secretion of the hormone from the sub-oesophageal ganglion neurosecretory cells (see p. 70). In any case these results suggest that, in the timing of the rhythmical cycle, oxidative processes are not of first importance in the hours preceding the activity peak. It is obvious that a great deal of information about the steps involved in a rhythmical process might come from careful studies of this sort.

In mammals there is evidence that the rhythms of at least some groups are very little affected by anaesthetics. The activity rhythm of the mouse *Peromyscus* is not measurably affected by either sodium pentabarbitol or ether anaesthesia acting over a period of about 6 h (Rawson, 1959, 1960).

## Metabolic Inhibitors

The effect on rhythmical processes of various enzyme inhibitors might be expected to give information about the clock, or the intermediary systems between the clock and the indicator processes. However so far even quite drastic measures have yielded very little in the way of results.

The circadian rhythm of spore discharge in the alga *Oedogonium* is unaffected in either phase or period by treatment with cyanide, sodium arsenate, 2:4 dinitrophenol or sodium fluoride, provided always that the concentration is not such that spore discharge is irreversibly prevented. Even when the concentration is such that discharge ceases altogether for several days when it restarts there is no change in the periodicity or phase of the rhythm. Other substances which leave the periodicity unaffected are iodoacetic acid, quinine, copper sulphate, cocaine, $\beta$-indoleacetic acid, A.T.P. and riboflavin (Bühnemann, 1955 *b*). The plant *Phaseolus* shows a similar insensitivity to such substances (Bunning, 1959).

The most comprehensive series of tests so far is that undertaken by Hastings (1960) on the rhythm of luminescence of *Gonyaulax*. Interpretation of the results have proved difficult because, although phase-shift has been recorded after application of some substances, the results were not always repeatable, and small phase-shifts occurred in some of the control experiments. Nevertheless the results are of considerable interest and may act as a stimulus for further work on these lines.

Addition of $5^1$-fluoro-$2^1$deoxyuridine, which affects DNA synthesis by blocking the conversion of deoxyuridine monophosphate to thymidine monophosphate, has no marked effect on the phase of the rhythm of luminescence. Addition of substances which block or accelerate protein synthesis similarly has little effect on the rhythm. Substances which stimulate oxygen consumption by uncoupling oxidative phosphorylation, for example 2:4 dinitrophenol and arsenate, have no affect on phase, with the exception of one sample in which a slight phase-shift was recorded. Two substances, mono- and dichlorophenyl dimethyl urea, which are very specific inhibitors of photosynthesis do not affect phase, and cells inhibited by these substances can still show a phase-shift if light is introduced at a new time of day. Other compounds which have no affect on phase-setting are gibberelin, kinetin and urethane.

Arsenite, an inhibitor of dithiol compounds, does cause a phase-shift, and the phase-shift is stable; the degree of phase-shift depends on the stage of the cycle of luminescence at which the arsenite is added.

Bruce and Pittendrigh (1960) have found that the length of the period of the phototactic rhythm of *Euglena* can be altered by culturing in the presence of deuterated water; the period alters to about 28 h after growth in this medium, but when the cells are transferred back to water and allowed to grow the period reverts to about 24 h.

### ACTION SPECTRA

Since light is the primary phase-setting factor for most circadian rhythms, studies of the action spectra for phase-setting, or phase-shifting, might yield valuable information about the action of light on any intracellular mechanisms involved

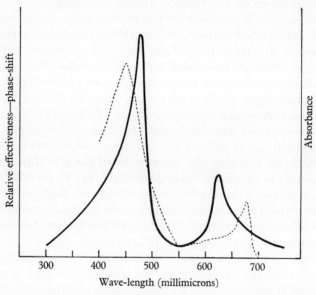

Fig. 25. The relative effectiveness of light of different wavelengths in producing a phase-shift in the rhythm of luminescence of *Gonyaulax* (unbroken line), and the absorption spectra of the total pigment content of *Gonyaulax* (dashed line). (Adapted from Ehret, 1960.)

in circadian rhythms. The selectivity of the intermediary sense organs would confuse the results of most such studies, but unicellular organisms lacking specialized sense organs have been the subject of two attempts to measure the action spectra.

Hastings and Sweeney (1960) have studied the effect of various wavelengths on the phase-setting of the rhythm of luminescence of *Gonyaulax*. The relative effectiveness of various wavelengths in achieving a phase-shift is shown in fig. 25. Similar studies have been made by Ehret (1959, 1960) on the rhythm of mating capacity of a strain of *Paramaecium bursaria* in which chlorophyl is absent. The results from the two studies show similarity in that the far red region has no effect on either organism, but in both phase-shifting is very effectively achieved by the ultra-violet region. The direction of phase-shift is not, however, the same in both species: ultra-violet always causes a delaying phase-shift in *Paramaecium*, regardless of when the exposure occurs relative to the circadian cycle, whereas in *Gonyaulax* the direction of phase-shift is always an advance. In both cases phase-shift is dose-dependent, and is photoreactivable by white light, an observation which suggests the implication of nucleic acid metabolism.

The efficiency of the 400 to 500 m$\mu$ wavelengths in causing phase-shifts in the rhythm of *Gonyaulax*, wavelengths which are for *Paramaecium* relatively inefficient, may be related to the presence of chlorophyll in the former, for the absorption spectrum of chlorophyll shows a large peak in this region. The action spectra for luminescence, photoenhancement and photoinhibition also suggest participation of both chlorophyll and peridinin. These action spectra correspond more closely with the absorption spectra of these pigments than does the action spectra for phase-shift.

As Hastings and Sweeney (1960) point out, it appears logical to conclude that different pigments which are physiologically active in various cellular processes may be utilized in different organisms as the photosensitizers for phase-shifting.

### X-irradiation Effects

In view of the efficiency of ultra-violet as a phasing factor it is interesting that X-irradiation has been shown to cause phase-shifts in the running activity of *Periplaneta* (Harker, 1958*a*). Irradiation doses of 100, 1000, and 10,000 r all cause phase-shifting, the magnitude of the delay being directly related to dose level. The phase-shift is persistent in constant darkness for as long as the rhythm is still measurable (the death rate is very high about two days after the highest dose has been given).

During the treatment, which occupies only a very short time, the cockroaches can move freely in a small chamber, so that there is no close-confinement effect. Doses of 100 r are followed by extreme hyperactivity, but about eight hours after treatment activity falls to a normal level. After irradiation doses of less than 100 r there is a phase-shift, but this is not stable, and the phases gradually move back to lie in their original relationship with solar time.

CHAPTER 5

# Physiological Processes Concerned with Diurnal Rhythms

## THE CELLULAR LEVEL

THERE seems to be little doubt that some clock-like cellular mechanism exists which is responsible for the observed physiological and biochemical rhythms. There is however no good example of an isolated and characteristic biochemical system with a rhythmical periodicity of about 24 h. On the other hand feedback mechanisms are known, for example the concentration of a particular compound may be regulated by means of effects upon enzyme synthesis, and it may be that short term feedback systems are involved in the observed physiological circadian rhythms.

The evidence concerning cellular rhythms (except in the case of unicellular organisms), let alone that of biochemical rhythms, is still fairly limited. Rhythmical changes in the volume of the nucleus of several kinds of plant tissue have been described by Bunning and Schöne-Schneiderhöhn (1957), and nuclear volume rhythms in the corpora allata of carabiid beetles have been measured by Klug (1958).

Halberg and his co-workers have made extensive studies of DNA and RNA metabolism of mammalian tissues, and have found significant circadian oscillations. In slices of mouse liver incubated with radioactive phosphate they found a two- to three-fold decrease in the relative specific activity of DNA (counts/min/$\mu$g of DNA P as % of counts/min/$\mu$g of acid soluble P) between 8 a.m. and midnight, and a sharp increase in activity at 4 a.m. In contrast to DNA activity the relative specific activity of both microsomal and nuclear RNA increased rapidly between 4 p.m. and 8 p.m. and then gradually dropped to the lower values which continued through the morning.

The periodicity in the relative specific activity of phospholipid

59

phosphorus is obliterated by adrenalectomy, and seems to depend largely upon the periodic secretion of adrenal hormones (Halberg, Halberg, Barnum and Bittner, 1959).

Circadian rhythms in nucleic acid metabolism have also been studied in a strain of non-dividing *Paramaecium* cells which nevertheless showed a circadian rhythm of mating capacity (Ehret, 1960). The nucleic acids were labelled by the addition of either adenine-8-$C^{14}$ or uridine-$C^{14}$, and three fractions were isolated by extraction with perchloric acid: acid soluble (nucleotide coenzymes, soluble RNA), RNA, and DNA. No major fluctuations in any of these fractions, in protein content, or in nucleic acid precursors in the form of adenine or uridine, were found in any of the light conditions used—alternating light and darkness, continuous darkness, or darkness interrupted by a short light period. However, small but statistically significant differences were found in the acid-soluble fraction on the day of a light interruption, and in RNA on the day following a light interruption. These differences are considered to be evidence of a circadian rhythm.

The DNA and RNA activity as measured by uptake of $P^{32}$ inorganic phosphate into non-dividing cells of *Gonyaulax*, showing circadian rhythms of bioluminescence, also show no pronounced fluctuations which can be associated with a circadian rhythm (Hastings, 1960). As Hastings points out, however, if there is any nucleic acid activity associated with rhythms it may be an extremely small percentage of the total nucleic acid of the cell, which would therefore be concealed in the results produced by the present methods.

Probably the most significant results concerning the biochemical mechanism come from the studies of Karakashian and Hastings (1962) on the effects of agents known to interfere with the synthesis of macromolecules. The rhythm of luminescence of *Gonyaulax* is abolished, after a lag period, by low concentrations of actinomycin D (0·02–0·08 $\mu$g/ml): at the lowest concentrations there is only partial inhibition of growth and the cells continue to be motile. Actinomycin also inhibits the rhythm of photosynthetic capacity although there is no immediate effect on the rate of photosynthesis. The sensitive period is 20–25 h before the maximum of the luminescent cycle being affected; the lag period is longer in the case of the photosynthetic rhythm.

Actinomycin is claimed to act by the inhibition of DNA-dependent synthesis of RNA, and it has been shown to inhibit a purified RNA polymerase in a cell-free system. In many cells protein synthesis depends on the continued production of a short-lived messenger RNA, and is therefore also sensitive to actinomycin. The inhibition of the rhythm of luminescence may therefore be due to either a direct effect of actinomycin on RNA synthesis, or to an indirect effect on protein synthesis. Karaka-shian and Hastings tried to distinguish between these two possibilities by measuring the effect of inhibitors of protein synthesis. Puromycin has a clear effect on the rhythm of luminescence and there is no apparent lag period, but the addition of chloramphenicol caused a striking increase in the amplitude of the rhythm. The authors suggest that this latter effect may be due to the stimulation of the production of messenger-like RNA, as has been reported to occur in some microbial systems in the presence of chloramphenicol.

Thus the question of the participation of RNA-directed protein synthesis is not settled by these studies; a stumbling block at present is our ignorance of the detailed biochemical mechanism of the action of either puromycin or chloramphenicol. In spite of this a close relation between the clock and RNA is indicated by these results, with or without a distinct link through protein synthesis.

In another series of experiments Sweeney and Haxo (1961) found that the direct participation of the nucleus appears not to be involved in the rhythm of photosynthetic capacity in *Acetabularia*, which persists for at least 5 cycles in enucleated plants. There seem to be two ways in which this result can be reconciled with those given above: (1) that there is a long-lived 'determinate' system, so that a rhythm may continue for a time in the absence of the nucleus; (2) that it is a cytoplasmic or chloroplast nucleoprotein component which controls the rhythm.

An interesting result has come from the analysis of variation in the concentration of the substrate and enzyme responsible for the luminescence of *Gonyaulax* (Hastings and Sweeney, 1957 b). Since the rhythm of luminescence is maintained in constant darkness it is unlikely that the circadian variation which has been found could be due to the direct action of light. As the authors point out the rapid flash of *Gonyaulax* is a very precisely controlled reaction and can be regarded as a type of cellular excitation. If the problem of the mechanism by which such an enzymatic reaction is controlled could be clarified we might be many steps closer to discovering the mechanism of the clock system.

## NERVOUS AND ENDOCRINE FACTORS IN INVERTEBRATES

Since hormones play such a major part in co-ordination, and in view of the ubiquity of physiological rhythms, it would be expected that hormonal secretion would commonly occur as a rhythmical phenomenon. If this were not so hormonal effects would override, or grossly upset, any circadian rhythms. That hormonal rhythms have been found is not, then, unexpected. On the other hand if hormones were rhythmically secreted then a great many physiological processes would also be expected to vary rhythmically, since they are radically affected by hormone concentration. These two alternatives underline the great dilemma of studies on circadian rhythms, the difficulty of separating cause and effect.

In one animal, the cockroach *Periplaneta*, an endocrine system has been found which not only participates in the production of a behavioural rhythm, that of running activity, but which can itself act as an autonomous 'clock' system (Harker, 1954, 1955, 1956, 1960 *a*, *b*, *c*). Since this is the only system so far discovered in multicellular organisms which can act as a timing mechanism when it is separated from its normal internal position in the body the evidence concerning the functioning of this endocrine clock, and that concerning its relationship with other systems in the cockroach, is described in some detail.

Cockroaches kept in a 12-h light:12-h darkness cycle show a very clear circadian rhythm of locomotor activity, in which the peak of activity occurs at the beginning of the dark period. The part played by hormones in producing this rhythm was first shown by means of experiments in which cockroaches were joined together in parabiosis (the blood systems being placed in continuity with one another). *Periplaneta*, unlike some other species of cockroach, ceases to show a measurable rhythm of activity when it has been in continuous light for a long period. When such an arrhythmic insect is joined parabiotically back to back with a cockroach which has been kept in 12 h light: 12 h darkness (this being done in such a manner that the only one of the pair free to walk is the arrhythmic one) the activity of this bottom insect, although it is still in continuous light, follows the same clear circadian rhythm which its upper partner was previously showing. The important point in this finding is that the rhythmical cockroach, taken from a cycling environment,

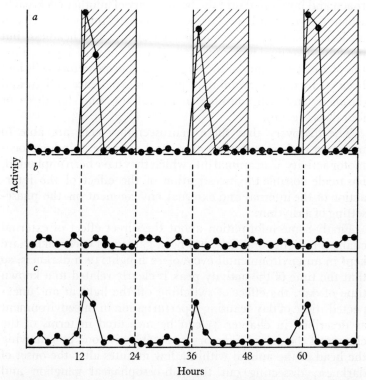

Fig. 26. (a) The activity of a cockroach (*Periplaneta*) in alternating light and darkness. (b) The activity of a headless cockroach in continuous light. (c) The activity of a headless cockroach into which the sub-oesophageal ganglion of the cockroach figured in (a) has been implanted.

imparts the *phase* of its activity rhythm to the lower animal; it therefore does not simply supply some substance which allows a sub-threshold rhythm of the bottom animal to become measurable. On the basis of these results it has been postulated that a secretion carried in either the blood or the tissues is involved in the production of the activity rhythm.

The source of the secretion has since been traced to the sub-oesophageal ganglion, and specifically to a group of neurosecretory cells lying on either side of the ganglion. Results from experiments in which different cells were cauterized in turn suggest that probably only four neurosecretory cells are involved.

The neurosecretory cells continue to secrete rhythmically when the sub-oesophageal ganglion is removed from a cockroach and implanted into the body cavity of another cockroach, which has previously had its head removed (the body cavity in an insect contains the open blood system). The headless, but implanted, animal now runs around at the time of day to which the phase of secretion of the implanted neurosecretory cells had been set by the donor animal's previous experience of light and darkness, and will continue to run at this time for three or four days (fig. 26).

The discovery that these neurosecretory cells are able to maintain their secretory rhythm, and are able to induce loco-motor activity in an animal into which they have been implanted, has made possible the investigation of the effect of the inter-action of the internal and external environment on the phase-setting of a rhythm.

Firstly some information about the direct effect of external changes in light conditions has been gained. If cockroaches are kept in an environmental cycle of 12 h light: 12 h darkness, so that the time of the activity peak is clearly related to a known time of day, the effect of switching off the light at an 'unexpected' time of day (causing a perturbation in the environment as described in chapter 4) can be measured in terms of the change of the actual secretory cycle. This is done by removing the head of the animal within a few minutes after the onset of darkness, dissecting out the sub-oesophageal ganglion and implanting this ganglion into the body of a headless animal. Fortunately it is relatively easy to remove the head of the animal whose ganglion is being tested in the dark, and the introduction of light, which is necessary for dissecting out the ganglion does not seem to act as a stimulus to a severed head. As was shown by the experiments already quoted, the time at which such an implanted animal runs about indicates the time at which the neurosecretory cells of the implant are secreting; therefore it is possible to measure any change in the timing of the secretory cycle which has been induced by the unexpected onset of darkness. Such experiments reveal that the neurosecretory cells react quite differently to a light-off stimulus at different times of day. It is possible to recognize three stages in the neurosecretory cycle; a stage when secretion will take place regardless of the external environmental conditions, a stage when secretion will take place if a light-off stimulus occurs ('possible' secretion stage), and a stage when secretion will not take place even under the stimulus of a change from light to darkness ('impossible' secretion stage) (fig. 27).

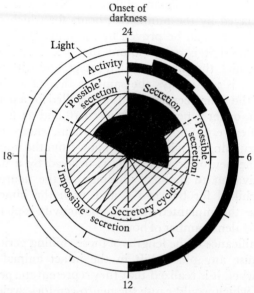

Fig. 27. Diagrammatic representation of the sub-oesophageal ganglion neurosecretory activity in *Periplaneta* in relation to the time of darkness and locomotory activity. (Harker, 1960*a*.)

These results are only obtained, however, if the sub-oesophageal ganglion is removed very soon after the light-off stimulus has been given; if the animal is left intact a quite different pattern of activity is observed. For example, if the stimulus is given at a time which corresponds to the stage in the neurosecretory cycle which has been termed 'impossible' secretion the next activity peak of the intact animal is phase-shifted so that it occurs several hours earlier than usual (an animal implanted with a ganglion which had been removed from an animal receiving an identical stimulation shows no phase-shift). These results suggest that internal factors can modify the immediate reaction of the neurosecretory cells to an environmental stimulus.

Evidence from another type of experiment suggests how the internal factors may play their part in phase-setting. It has already been mentioned (p. 51) that the 'clock' appears to stop at low temperatures. A method has been devised (Brown and Harker, 1960) which enables the neurosecretory cells to be chilled *in situ* while the rest of the body is kept at room temperature. When the cells are chilled in this way the secretory cycle is delayed by the number of hours for which chilling is maintained. This can be proved by the implantation test. On the other hand it is again found that the behaviour of an animal in which the ganglion is left intact, and with its normal connections, is not always the same as that of an implanted animal. If the neurosecretory cells are chilled for a period of up to 4 to 5 h, or for a period of between 18 and 24 h, an intact animal, after return to normal conditions, shows no phase-shift in its running activity (yet we know from the implantation test that the neurosecretory cycle is delayed by a number of hours equivalent to those of chilling). If the neurosecretory cells are chilled for between 5 and 17 h the intact animal behaves like the implanted animal, and activity is phase-shifted by the equivalent number of hours.

The significance of the lengths of those chilling periods which do not cause any phase-shift in the intact animal becomes apparent when it is realized that they represent the periods in a 24-h cycle which would cause the neurosecretory cycle to move out of phase with any other 24-h cycle by only $\pm 5$ to 6 h, and that the neurosecretory cycle is known to have a 'possible' secretory stage of 5 to 6 h both before and after its normal secretory stage. This means that any 24-h cyclical process which is not affected by the chilling, and which could stimulate the sub-oesophageal ganglion cells to secrete, would be effective at $\pm 5$ to 6 h from the normal time of secretion (fig. 28). If the unaffected 24-h cyclical process acted as a stimulus to the neurosecretory cells at other times in the 24-h cycle there would be no response because the stimulus would fall within the 'impossible' secretion stage. Therefore the results strongly suggest the presence of a second 24-h cyclical process which is maintained outside the sub-oesophageal ganglion.

Since it is known that a light-off stimulus can set the phase of the cockroach activity rhythm further experiments have been carried out in which a light-off stimulus has been given to animals while the neurosecretory cells are actually being chilled. If a rhythmical process which is controlled by some region outside the sub-oesophageal ganglion persists during chilling, as is

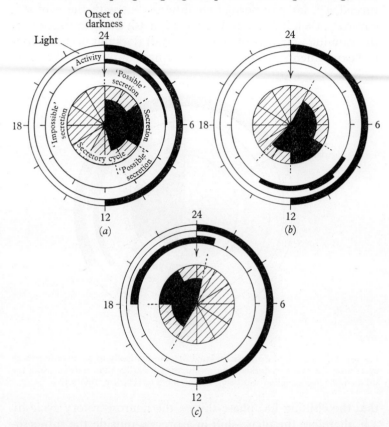

Fig. 28. The activity of *Periplaneta* related diagrammatically to the neurosecretory cycle after the sub-oesophageal ganglion has been chilled for (*a*) 4 h, (*b*) for 8 h, (*c*) for 18 h. (Harker, 1960 *b*).

so strongly suggested by the previous results, and if this process should respond to a light-off stimulus, then it would be possible to phase-shift this process so that its phases lie in a new, but known relationship to the neurosecretory cycle; the subsequent activity rhythm of the animal would give further information about the dual control system. It has proved possible to perform such experiments by simulating a light-off stimulus by painting black the compound eyes and ocelli of the cockroach. If the chilling of the neurosecretory cells starts 4 h before the normal time of running activity, the eyes painted over 15 min after the temperature has been lowered, and the animal returned 4 h later to normal conditions (but with the eyes still covered), then it is found that there is a 4-h delay in the time of the running activity. Such a phase-shift does not occur in control experiments in which the eyes remain unpainted. We know

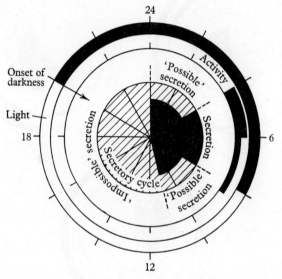

Fig. 29. The activity of *Periplaneta* related diagrammatically to the neurosecretory cycle after the onset of darkness has occurred 4 h earlier than normal while the sub-oesophageal ganglion was being chilled for 4 h. (Harker, 1960 *b*.)

that the chilling has phase-shifted the neurosecretory cycle by 4 h, therefore the phase-shift in a process outside the sub-oesophageal ganglion, which must have occurred to give these results, will bring the time at which this process stimulates the neurosecretory cells to a time when the 'impossible' secretion stage is operating—therefore neurosecretion will occur 4 h later than normal, as it does in the implantation test (fig. 29).

In another group of experiments ganglia were chilled for 8 h, beginning just before the time of the activity peak, and the eyes blackened 4 h after the beginning of the chilling treatment.

The 8-h chilling delays the secretory cycle for 8 h, but the light-off stimulus, coming 4 h later than the normal time of activity, causes a phase-shift in the second rhythmical cycle so that it stimulates the neurosecretory cells 4 h later than usual; since this coincides with the 'possible' secretion stage the subsequent running activity is delayed by only 4 h instead of the 8 h delay which is shown by the unpainted controls (fig. 30).

It appears, therefore, that there are at least two 'clock' mechanisms involved in the production of the rhythm of running activity, each affected by changes from light to darkness,

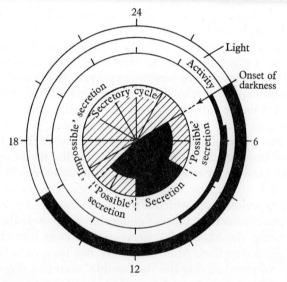

Fig. 30. The activity of *Periplaneta* related diagrammatically to the neurosecretory cycle after the time of onset of darkness had occurred 4 h later than normal, while the sub-oesophageal ganglion was being chilled for 8 h. (Harker, 1960b.)

but that the neurosecretory cycle is reset via the cycle maintained by some 'clock' outside the sub-oesophageal ganglion (master clock), and that immediate resetting is only achieved when the stimulus given by this latter cycle occurs within about 5 h on either side of the normal secretory stage. Yet the neurosecretory cycle is maintained autonomously when the cells are removed from the influence of the master clock, and the master clock itself is also autonomous since its cycle is maintained throughout the chilling procedure.

The interaction of the two cycles explains many of the results obtained when a cockroach is given a light-off stimulus at an 'unexpected' time of day, but they do not explain the type of phase-shift (described on p. 38) which occurs after a sudden dark stimulus is given during the time of the 'impossible' secretion stage. After such stimulation the next episode of secretion is brought forward by 4 to 6 h. If the sub-oesophageal ganglion is removed any time up to 8 h after this dark stimulus, the implantation test shows that there has in fact been no phase-shift in the time of neurosecretion. It is possible that the master clock cycle described above, which will have been reset by the stimulus, may affect the neurosecretory cycle in some way, although, as has been shown, it is unable to reset the neurosecretory cycle immediately. It is also possible that yet some other cycle is involved, or that some other process which ultimately affects the neurosecretory cycle is affected by the dark stimulus. There is some evidence that the latter is the case, and that another substance is involved in the maintenance of the secretion in the sub-oesophageal ganglion.

The first piece of evidence for this view comes from the fact that the implanted sub-oesophageal ganglia maintain their secretory activity for only a few days, although the cells appear to be still healthy. The second piece of evidence comes from a series of experiments in which it was found that a secretion from the corpus cardiacum (another endocrine organ near the brain) passes along the axons of a nerve which runs into the sub-oesophageal ganglion. After this nerve has been cut a light-off stimulus received during the 'impossible' secretion stage of the neurosecretory cycle is no longer followed by a phase-shift in the running activity rhythm; furthermore several days after nerve severance the secretory rhythm of the cells disappears. The secretory rhythm is even lost when animals in which the nerve has been cut are kept in alternating light and darkness; it would therefore appear that an integral part of the timing mechanism has been removed by this operation. The fact that animals do retain an activity rhythm for a few days after nerve severance, and that implanted ganglia also retain their rhythm for a similar period may mean that there is always enough corpus cardiacum hormone present in the ganglion to support the secretory rhythm for a time, and that the fading of the rhythm occurs only when this substance is exhausted.

There is no evidence that the neurosecretory 'clock' has stopped, only that it is no longer effective in producing an activity rhythm.

Further experiments have provided evidence that the neurosecretory clock can indeed continue to measure time even when no secretion, or at least an ineffective concentration of secretion, appears to be released from the neurosecretory cells. When a sub-oesophageal ganglion is implanted into a normal cockroach which has been living in the same environment as the donor of the ganglion (so that the neurosecretory cells of the implant (B) and those of the recipient (A) will secrete at the same time of day), the running activity of the animal into which the implantation is made shows a decrease in amplitude. If this animal's own ganglion (A) is dissected out 24 h after the extra ganglion (B) has been implanted it is found, by the 'implantation' test, that ganglion (A) no longer shows a rhythm: the original implant (B) however continues to secrete rhythmically. It seems that under the influence of an implanted ganglion the secretory activity of the animal's own ganglion is inhibited, provided that the two are secreting in phase with each other. If the implant is left in position for only 24 h, and then removed, the neurosecretory cycle of the animal's own ganglion reappears about 72 h after the time of the removal of the implant. The significant feature of these results is that, provided that there has been no change in the environmental light conditions, the rhythm reappears with the same phase-setting as it was showing prior to inhibition. The clock, therefore, may still be running even when it is producing no observable effect on the locomotor activity. The inhibitory effect of high hormone concentration resulting from the presence of two ganglia secreting at the same time of day appears to be regulated through the nervous connections of the sub-oesophageal ganglion with the corpora allata, and through the nervous connection of the corpora allata with the recurrent nerve (Harker, 1960 c).

All the evidence quoted above shows that some cells in a complex animal can maintain a circadian rhythm when they are isolated from the other tissues of the body, but that the activity of these cells is normally under control of at least two other systems, one of which is itself able to maintain a circadian rhythm independently of the cells which it controls. These results do not encourage the expectation that there will be a simple answer to the problem of the whereabouts and the functioning of a mechanism controlling all circadian rhythms.

A somewhat similar endocrine effect has been found in the control of the chromatophore rhythm of the crab *Carcinus* (Powell in Bliss, 1962). In this case removal of the eyestalks, which in crabs contain many neurosecretory cells and are a known source of hormones controlling chromatophore movement, causes loss of the chromatophore rhythm. When eyestalks, taken from crabs which have been in a 6-h light:6-h darkness cycle, are implanted into eyestalkless arrhythmic crabs the rhythm reappears with a phase-setting correlated with the 6-h light:6-h darkness cycle. This is a most unexpected result since it is seldom that a 12-h cycle can be induced in a process normally showing a circadian rhythm (see p. 35), and in this case it appears not only to have been induced, but also to be maintained by an implant.

A possible participation of the neuroendocrine system in the locomotor activity rhythm of the crab *Gecarcinus* is suggested by Bliss (1962). It is possible that an activity rhythm persists after eyestalk removal, although only a small amount of evidence is so far available, but it seems likely that quite a large change in period length occurs; it is of course possible that the change in the free-running period is due to a change in the whole physiological state of the animal which may follow eyestalk removal.

### NERVOUS AND ENDOCRINE FACTORS IN VERTEBRATES

The great majority of the records of physiological, as opposed to behavioural, rhythms come from studies made on mammals. This is not because more processes are rhythmical in mammals, but because more physiological studies are made on this class. By no means all of the known rhythms can be described here, but an attempt is made to discuss the interrelations of various rhythms, and the part played by the endocrine and nervous systems in their control.

### *Adrenal Control*

The blood level of corticosterone in mice shows a circadian periodicity in which the peak blood level occurs about 10 h after the beginning of the light period in a 12-h light:12-h darkness cycle. Furthermore there are periodic changes in the hormone content of the adrenal cortex, and a mitotic rhythm is also present. The peak of mitosis occurs at the time of minimal corticosterone concentration (Halberg, Peterson and Silber, 1959) (fig. 31).

At least three other rhythms are known to be related to the functioning of the adrenals, the rhythmical variation in the number of eosinophils, the mitotic rhythms in the epidermis, and the rhythm of phospholipid concentration in the liver. The period of each of these related rhythms changes under conditions of adrenal insufficiency (fig. 32). Adrenalectomy eliminates the mitotic cycle (Halberg, Halberg *et al.*, 1959; Bullough

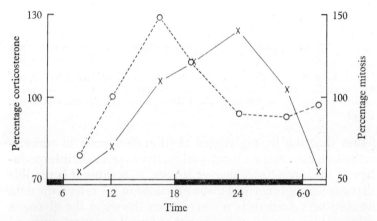

Fig. 31. Rhythm of corticosterone in the adrenal of mice (O – – – O) and the mitotic rhythm in the adrenal cortex (× ——×). (After Halberg, Halberg, Barnum and Bittner, 1959.)

and Lawrence, 1961); but the rhythm is not depressed by high doses of adrenaline (Bullough, 1962). In addition, the sensitivity of the tissue to adrenaline may vary from time to time, and in periods of low sensitivity there is no sign of a mitotic rhythm. Despite these findings it seems that the mitotic rhythm is related to the high rate of adrenaline secretion when the animal is asleep. It is not surprising, therefore, to find that there is a considerable degree of synchronization between the phases of the mitotic rhythms of various tissues and organs; for example those of the skin epidermis and the liver parenchyma in mice (Halberg, 1959), and the pinnal epidermis, the oral mucosa, and

73

the connective tissue in the periodontal membrane of the rat (Halberg, Zander, Houghlam and Mühlemann, 1954). Yet out of phase with these mitotic rhythms is the rhythm of mitosis in the adrenals themselves, which poses the question as to what regulates the rhythm in the adrenals.

Another type of rhythm which may be associated with the adrenals is that of carbohydrate metabolism. Agren (1935)

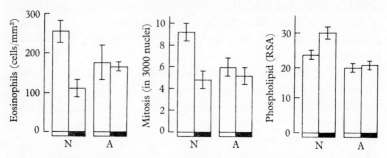

Fig. 32. Effect of adrenalectomy on the blood eosinophils, epidermal mitosis, and the phospholipid metabolism in the mouse. N = normal animals, A = adrenalectomized animals. ☐ day ▓ night. (After Halberg, Halberg, Barnum and Bittner, 1959.)

found that the liver glycogen rhythm disappears in adrenalectomized rats. Some difficulty arises, however, in the interpretation of these results (Solberger, 1955 a, b) because it is possible that some ectopic cortical tissue remained in the rats: if the data from suspect animals is removed from the series the glycogen values which remain are so low that for all practical purposes glycogen seems to have disappeared from adrenalectomized rat liver (Halberg, 1959). The persistence of glycogen rhythm in the liver of starved animals suggests that feeding is not the controlling factor in this periodicity, but confirmation of the results has not been given by all workers, and it has been suggested that there may be two peaks, one of which is related to feeding while the other is independent of it (Sjoegren, Nordenskjoeld, Holmgren and Mollerström, 1938).

The adrenal medulla, as well as the cortex, shows some rhythmicity, and epinephrine concentration in the adrenal, as well as in the blood, exhibits a circadian rhythm.

Not all rhythms are abolished by adrenalectomy, for instance that of the iron content of human serum is maintained in such a condition (Howard, 1952).

*Hypophysis and Hypothalamus*

In mice both the rhythm of epithelial mitosis and rectal temperature are retained after hypophysectomy, as is the activity rhythm of rats (Levinson, Welsh and Abramowitz, 1941). The feeding rhythms of mice is however obliterated after hypothalamic lesion or other treatments which damage the hypothalamus (Anlicker and Mayer, 1956; Skinner, 1957).

Two of the most striking rhythms shown are those of locomotor activity and of sleep and wakefulness; these are closely related but are by no means always synonomous. We are as yet far from understanding the complex of interacting factors involved in the states of sleep and wakefulness, and the subject is one of the most complex in the field of mammalian physiology. Because of their possible relevance to the internal clock problem a brief discussion of these states would seem to be justified.

Hibernating animals (which wake at a time of day determined before hibernation by the usual environmental phase-setters for rhythms) are eminently suitable subjects for consideration. In such animals the role of the sympathetic nervous system in the arousal from hibernation seems to be definitely established. The centres of sleep and wakefulness are located in the hypothalamus: that of wakefulness being situated in the posterior hypothalamus, probably being part of the sympathetic hypothalamic centre (Suomalainen, 1961).

The part played by the brain in sleep is very complex and, as said before, by no means fully understood. The major provision seems to be reticular deactivation, and this may be brought about by both ascending and descending inhibitory effects from the bulbar region. The ascending effects act at either, or both, the mesencephalic reticular level and the thalamic level. The descending effect comes from an aroused cortex (Dell, 1961). In addition active inhibition of the ascending reticular system may result in a slackening of the activating influence of the brain stem (Moruzzi, 1961).

It is of particular interest, since light is a strong phase-setting factor, that some fibres of the optic nerve go to the hypothalamus and the reticular formation, and also that various peripheral afferent stimuli (visual, auditory and somatic) can be conducted corticipetally through the central brain stem (Russell, 1957; Startzl, Taylor and Magoun, 1951; Magoun, 1952). Effects of such excitation lead to activation of the cerebral cortex (French, Verseano and Magoun, 1953) and arousal of the animal. On the other hand the cerebral cortex does not appear to be essential for continuance of circadian rhythms (Kleitman, 1952; Rothmann, 1923). Furthermore temperature and eosinophil rhythms persist in paranoid and catatonic schizophrenic patients after regression induced by intensive electroshock therapy (Fleeson, Glueck and Halberg, 1957). Halberg and his associates (1959) point out that in such a state the patients are mute, disorientated, due to complex memory loss, are doubly incontinent, and show neurological changes associated with extrapyramidal tract activity: despite this, the rhythms persist.

A neural mechanism exhibiting 24-h periodicity of sensitivity has been suggested as the controlling factor in the release of ovulation-inducing hormone in rats (Everett and Sawyer, 1950). In another class of vertebrates, birds, there is further evidence for this: injection of estradiol benzoate into regularly ovulating hens suppresses ovulation for some hours, but the next day ovulation occurs at the normal time, regardless of when the inhibition is removed. Fraps (1954) concludes that the neural mechanism controlling the hormone inducing ovulation follows a circadian periodicity in its response to excitatory hormones.

## DEVELOPMENT OF CIRCADIAN RHYTHMS
### IN YOUNG VERTEBRATES

The time at which rhythms first appear in developing vertebrates might give some indication of the controlling mechanism involved in rhythms, although as in all problems of the controlling mechanisms of rhythms there is great difficulty in distinguishing between the controlling mechanism and the chain of events occurring between the 'clock' and the indicator process.

The embryos of birds show circadian rhythms while still in the egg: if the egg is kept in a light:darkness cycle the embryo follows a rhythm of movement (Heibel, 1949). An activity rhythm is also apparent immediately after hatching (Aschoff, 1953), as is a rhythm of oxygen consumption (Bacq, 1929; Burckard, Dontcheff and Kayser, 1933; Barrott, Fritz, Pringle and Titus, 1938). Both these rhythms are persistent in constant conditions. The rhythm of body temperature however does not appear until about nine days after hatching (Baldwin and Kendeigh, 1932).

Lizards hatching in a constant environment show clear rhythms of running activity, individuals varying slightly in their periodicity (Hoffmann, 1957).

Rats and mice show evidence of rhythmical functions from birth (Stier, 1930, 1933; Wolf, 1930). It is difficult to estimate the effect which the parents may be having on the young in mammals, but Folk (1957) has designed experiments in which any information about periodicity passed on from the mother during suckling would be obviated. He changed young mice round between foster-mothers kept in environmental cycles with different phase-settings: the young mice still exhibited circadian rhythms. It is of course still possible that information about periodicity is passed on to the embryo from the mother before birth.

No evidence of periodicity has been found in the human embryo. In particular there is a lack of periodicity in the heart-beat although this shows a distinct periodicity in the mother, and is an easily measured factor in the embryo (Hellbrügge, 1960). A long and detailed study of the time of development of rhythms in human infants has been made by Hellbrügge. He has shown that rhythms of different functions develop at different times after birth: that of electrical skin resistance develops in the first week, that of total urine excretion in the second or third week, although rhythmical distribution of potassium and sodium excretion does not appear until the end of the second month (the first depends on the functioning of the glomeruli, the second on the functioning of the tubuli), and the rhythm of heart-beat rate develops in the sixth week. A rhythm of body temperature begins about the third week (Jundell, 1904).

Development of sleep rhythms have been studied many times; a monophasic rhythm in sleep and waking takes place about the sixteenth week, even in children on a self-demand feeding schedule (Kleitman and Engelmann, 1953).

It is perhaps significant that the rhythm of sleep and pulse-frequency develop later in premature children than in those born at term; this suggests that the time for development of these rhythms is not due to a required period spent in a synchronizing environment, but requires a certain stage in development to be reached.

CHAPTER 6

# Abnormalities in Rhythmical Systems

DISPHASIA

STUDIES made of several different rhythmical processes within a single animal clearly reveal a close interrelationship between the timing of the phases of the different physiological rhythms. Dissociation of the phases of two or more rhythms may occur, particularly when environmental stimuli provoke phase-shifting, but such dissociation has never proved to be more than a short term phenomenon: mutual entrainment, and (or) entrainment by the environment, cause the quite rapid return of the phases to their normal relationship. Indeed the relationship between the phases of the rhythms occurring in the one individual appears to be so stable that any treatment which forced the dissociation of the phases for any length of time might be expected to produce serious physiological stress.

It has not, so far, proved possible to achieve such a forced dissociation, but the discovery of the autonomous secretory rhythm of the sub-oesophageal ganglion neurosecretory cells in the cockroach (p. 64) has made it possible to produce a dual secretory cycle in the one individual, thereby introducing one secretory cycle which must be 12 h out of phase with any other rhythmical function which is tied to a 24-h periodicity.

The method of producing the dual cycle is fairly simple. In *Periplaneta* the time of secretion from the neurosecretory cells is determined, in a light:darkness cycle, by the time of the beginning of darkness: the phase, once determined, is stable. Therefore when two groups of cockroaches are kept in light:darkness cycles which are 12 h out of phase with each other, and the sub-oesophageal ganglia from one group are implanted into the cockroaches of the other group, the implanted animals will contain two ganglia secreting 12 h out of phase with each other. If the implants are renewed every day for at least four days the recipient cockroaches begin to show signs of serious pathological

78

disorder, and by sixteen days from the beginning of the experiment malignant tumours are found in practically all animals. The tumours mainly occur in the midgut, but they are occasionally found in the foregut. The site of tumour formation is not related to the position of the implant. The tumours so produced are transplantable and metastasize (Harker, 1958b).

The normal midgut of the cockroach contains regularly arranged nidi of regenerative cells in which mitosis appears to occur at a fairly constant rate. It is these cells which are, at least initially, most affected by the presence of the implanted sub-oesophageal ganglion, the number of mitoses more than doubling after only two days of treatment. The cells very soon appear to move out of their normal position, and to migrate into the adjacent connective tissue layer and, in some cases, into the haemocoele. It is these migratory cells which appear to form the tumours.

TABLE I

Midgut of treated animals

|  | Normal midgut | 2 days after implantation | 4 days after implantation | 8 days after implantation |
|---|---|---|---|---|
| Mitoses per 25 nidi, means of 10 counts | 8·1 | 18·0 | 18·3 | Nidi not clearly defined |
| Cells per nidus, means of 50 counts | 8 (8–10) | 16 (16–18) | 16 (16–18) | Nidi not clearly defined |
| No. epithelial cells between nidi, means of 50 counts | 16 | 28 | 30 | 30 |

In these experiments the high hormone concentration produced as a result of the presence of two ganglia, rather than any rhythmical phenomenon, could be the critical factor in tumour production. A high concentration of hormone, however, does not have any effect when it is produced by the implantation of even a large number of ganglia, providing that these are secreting in phase with the animal's own ganglion. Neither does injection of ganglionic extracts, when this is done close to the time of normal secretion, result in the production of tumours. Extracts injected 12 h after the active phase, the injections being repeated for at least five days, do however have an effect on the recipients, and tumours are formed in some animals.

79

There is some evidence that, when the environmental cycle has been suddenly phase-shifted through 12 h, different processes in cockroaches phase-shift at different rates. It is, therefore, of interest that when sub-oesophageal ganglia taken from animals living in normal conditions are implanted into animals which have been kept in light at night and darkness during the day, for just long enough for the running activity to have phase-shifted to the new position, the implantation has to be carried on for twice as long as in the previous experiments if tumours are to be induced. The percentage of tumours formed in such conditions also drops considerably. It seems likely that either the cells of the gut, or some other cells involved in the abnormal growth, show a rhythm of sensitivity to the sub-oesophageal ganglion hormone, and that this rhythm has not been phase-shifted as rapidly as the running activity rhythm.

Tumours induced by any of the above treatments can be, to some extent, controlled by the same hormone as that which is initially responsible for their production. If daily implantations are made into animals in which tumours are present, using sub-oesophageal ganglia from cockroaches which are secreting at the same time as that of the recipient, then the tumours cease to develop, and the number of mitoses drops to normal. As soon as this treatment ceases the growth of the tumours continues, showing that the cells have undergone an irreversible change.

It is not possible to induce persistent secretion by an animal's own ganglion at 12-hourly intervals, but when cockroaches are kept in a 2-h light: 6 h darkness cycle our knowledge of the way in which phase-shifts are induced in the secretory cycle (p. 65) would lead to the expectation that constant resetting might occur in such a cycle for at least some time. When animals are kept in this environmental cycle many show gross pathological changes in the gut (Harker, 1961); these changes have not been studied in detail and there is as yet no evidence of malignancy.

Pittendrigh (1960) has begun a study on the comparative effect of periodic and aperiodic environmental conditions on the expression in *Drosophila* of genes of variable penetrance. One of the stocks used has a recessive allele responsible for melanotic pseudotumours, and when this stock was kept in constant light the penetrance of the gene fell from 90 per cent to 20 per cent by the fifth generation. From the brief published description of the experiments it is difficult to assess the results, and in view of both the difficulty in determining the status of melanotic tumours in *Drosophila*, and, in particular, the very marked effect on the incidence of melanotic tumours of the culture medium (which may itself be affected by light) (Harker, 1963), much more evidence is needed before any conclusions can be drawn.

## SYNCHRONIZATION OF CYCLES IN DISEASED SYSTEMS

Richter (1960) has drawn attention to the many cyclical phenomena which become manifest in man under pathological conditions. The majority of these cycles have considerably

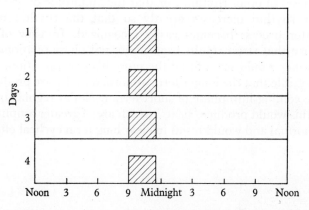

Fig. 33. Record of a woman suffering from Parkinsonism, the shaded areas represent the times at which she was able to feed, walk and talk normally. (After Richter, 1960.)

longer periods than 24 h, and are therefore not discussed here in detail, but there are some recorded cases in which the period of the abnormal cycles is almost precisely 24 h. One case cited is that of a 28-year-old woman suffering from Parkinsonism: during a nine-year observation period in hospital the patient was, from near midnight till 9 p.m. the following night, unable to speak clearly, walk or feed herself because of the rigidity and tremors of her arms and legs. Quite suddenly near 9 p.m. each evening the rigidity and tremors disappeared and she was able to walk, talk clearly and look after herself (fig. 33).

Richter has suggested that the cells or tissues of every organ in man have an inherent cycle which is characteristic of the organ: in normal conditions the cycles of each of the units making up an organ are out of phase with each other, but under the stress of shock, trauma or allergy these cycles are brought into phase with each other so that the inherent cycle of the organ as a whole can be recognized. The evidence presented throughout the earlier chapters of this book suggests strongly that there is a fixed relationship between the phases of the various physiological circadian rhythms, and it seems unlikely that the cellular rhythms should all be randomly out of phase with each other. It may, however, be that under stress conditions either the phase-relationships between different processes alter, perhaps because, the dominance of one particular rhythm which entrains the phases of other rhythms, or that the amplitude of one particular rhythm increases greatly so that the rhythm of an indicator process becomes easily recognized. In view of the evidence that stress results from abnormal phase-relationships the former would seem to be the more likely supposition. It is also possible that the longer term abnormal cycles are produced by the coming into phase of short term cycles at regular intervals: this would produce 'beats' which are of greater amplitude than normal and would result in clear long term cyclical effects.

## CHAPTER 7

# External Time-Signals?

THE term 'internal clock' which has been mentioned many times in this book carries with it the implication that the ultimate controlling system of circadian rhythms is to be found as a biochemical or biophysical system within the organism. At the present time we are still not clear how this system works, where it is situated, or what reactions are involved. The hypothesis that there is such a system is based largely on the fact that the observed rhythms are maintained apparently independently of environmental time-signals, and that in any one type of constant environmental condition individuals of the same species show significant variation in their periodicity.

The possibility that external time-signals reach the organism has, however, been strongly supported by one group of investigators led by F. A. Brown. Brown and his colleagues have questioned the assumption that in controlled 'constant' conditions an organism is completely isolated from every environmental factor to which it might be sensitive, and in a long series of papers this group has presented evidence from which they conclude that organisms under 'constant' laboratory conditions do derive information about atmospheric rhythms.

Part of Brown's evidence comes from consideration of the colour-change rhythm in the crab *Uca*, which is said to show not only a diurnal rhythm but also a tidal rhythm (Brown, Sandeen and Fingerman, 1952), and a semi-lunar rhythm (Brown, Webb and Graves, 1952). Measurements of oxygen consumption have also been made over long periods of time, and again statistical analyses reveal diurnal, tidal and semi-lunar rhythms (Brown, Bennett and Webb, 1953, 1954). The type of statistical analysis used involves averaging the corresponding hourly data of several days and 'slipping' the data one hour per day in order to remove the obscuring effect of the interaction of a lunar rhythm and the diurnal rhythm. When this procedure is used the daily oxygen consumption shows a monthly variation in which the time of the peaks of uptake in the two-week period straddling a new moon is different from the time at which the peaks occur in the two-week period straddling a full moon.

Correlation has also been shown between the oxygen consumption of *Uca* (and many other organisms ranging from slices of potato tissue to salamanders) and the external barometric pressure, even when the organisms are kept in pressure- and temperature-controlled chambers (fig. 34). The rate of

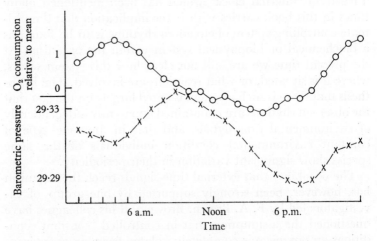

Fig. 34. The mean daily metabolic cycle of *Uca* relative to the barometric pressure (O = oxygen consumption, × = barometric pressure). (After Brown, 1960.)

change of barometric pressure external to the controlled environment appears to be related to the rate of oxygen consumption (Brown, Freeland and Ralph, 1955; Brown, Webb, Bennett and Sandeen, 1955). However the correlation is sometimes significant only for the afternoon peak of oxygen uptake and sometimes only for the morning peak; at times there is no correlation between the peaks and the barometric pressure fluctuations, but at such times the mean oxygen uptake over 24 h can be shown to be correlated with pressure. In all cases the correlation is said to be in respect of barometric pressure of from one to seven days *ahead* rather than that of the day of the experiment. With this number of qualifications it is difficult to be certain that the correlation does not represent chance coincidences of two independent 24-h periodicities.

In the specific case of the potato Brown has shown that the mean daily cycle of oxygen consumption, taken from the results of a three-year study, shows low metabolic rate at night, a rapid rise during the early morning and the occurrence of three maxima during the day (fig. 35 a). The percentage increase

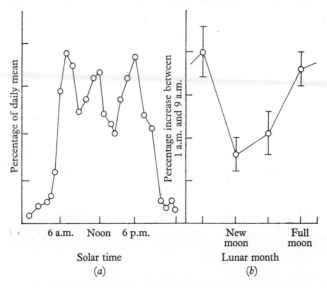

Fig. 35. (a) The mean daily cycle of oxygen consumption of potatoes for a continuous three-year study. (b) The mean lunar monthly cycle modulation in the form of the daily cycle. (After Brown, 1960.)

from the night low point (at 1 a.m.) to the morning peak at 9 a.m. varies over the lunar month, with the smallest value near the time of the new moon and the largest near the time of the third quarter of the moon (fig. 35 b). The relationship between the external barometric pressure and oxygen consumption can be seen if the form of the metabolic cycle in summer and that in winter are compared with the pressure cycle of these two periods. During the summer the 6 a.m. metabolic value is directly related to the mean rate of pressure change between 2 a.m. and 6 a.m., and the 6 p.m. metabolic value is inversely related to the mean rate of pressure change from 2 p.m. to 6 p.m. In winter, however, the morning metabolic value becomes inversely related to the pressure change. The pressure curve at these two times of year varies in that during the critical afternoon period the summer curve shows a decrease in pressure while the winter curve shows an increase (fig. 36).

The statistical treatment of the data quoted in these experiments has been criticized by a number of authors, and particularly by Cole (1957). Cole treated numbers taken from a table of random numbers in the way in which oxygen consumption measurements have been treated in the above experiments,

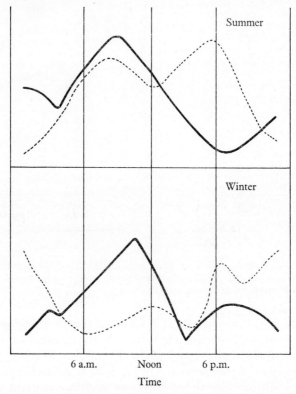

Fig. 36. The mean monthly forms of the daily cycle of barometric pressure for a sample June and December (unbroken line), and the average form of the daily metabolic cycle for potatoes in summer and winter (dashed line). (After Brown, 1960.)

and plotted them against successive hours of the day. Before the data had been 'slipped' an hour each day no rhythm was apparent, but after slipping a clear 'daily' rhythm emerged. Cole also used the standard procedure of smoothing the curves by means of a 3-point moving average, and again a clear 'daily' rhythm was produced with a peak at 3 a.m. and a minimum 12 h later. He further noted that 'like any other biological clock this was independent of the temperature at which the observations were made'.

However, even if the method of treating the figures in itself gives rise to an apparent daily, or lunar, rhythmicity it is difficult to see how the statistical treatment produces changes in sign at different times of the year (fig. 37).

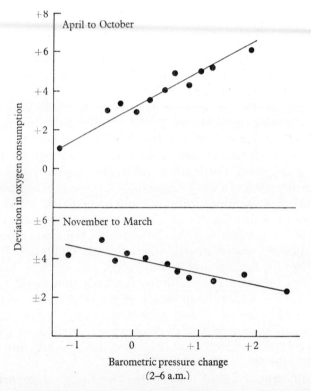

Fig. 37. The regressional relations between morning pressure change and morning metabolism during the two portions of the year. (After Brown, 1960.)

In searching for environmental variables which are not controlled in the normal laboratory constant conditions Brown has found some correlation between cosmic ray cycles and metabolic cycles (Brown, Webb and Bennett, 1958). Other authors have investigated the effect of screening organisms by thick lead plates calculated to screen the organism from about 70 per cent of the normal cosmic radiation: the eclosion of *Drosophila* does not appear to be affected by such screening (Pittendrigh, 1958).

Terrestrial magnetism is known to fluctuate rhythmically with solar and lunar periods. The directional movement of the mud snail *Nassarius* has been measured at different times of day after the snails have crept out of a funnel-shaped enclosure in which the exit is orientated towards magnetic south. The average path of orientation is found to be related to time of day and also to be affected by a lunar day cycle. Increasing the strength of the magnetic field experimentally, so that it is nine times that of the earth's field but similarly orientated, increases the magnitude of the response and does not alter the daily cycle. When a second magnetic field was introduced and rotated so that the lines of force were at right angles to the long axis of the snail's body, the first field being kept parallel to the snail's long axis, the snails orientated relative to the parallel field during the middle of the day and relative to the right angle field at night: the angle of orientation could also be correlated with the hour angle of the moon. Furthermore it has been shown that on the average the effect of a weak magnetic field during the day-time is to induce right-handed turning of snails immediately before and after both new and full moons, and maximum left-handed turning between the first and third quarters of the moon (Brown, Brett, Bennett and Barnwell, 1960; Brown, Webb and Brett, 1960; Brown, Bennett and Webb, 1960).

In view of these results Brown contends that at all times organisms are under the influence of environmental variables which could give time-cues. This contention is by no means accepted by the majority of those who work on rhythms, but so far no evidence whatever has been recorded which in any way disproves the effects observed by Brown.

A different type of experiment involving transport of organisms, kept in constant environment containers, through long distances in an eastwards or westwards direction has also given results which Brown and his colleagues claim to indicate the presence of an external controlling system. Oysters collected in New Haven, Connecticut, were transported to Evanston, Illinois, and the time of opening of the valves observed. Immediately after arrival at the new location the activity of the oysters appeared to be phased to the tide times of their place of origin, but after two weeks the time of opening of the oysters had phase-shifted to correspond with the lunar cycle in Evanston; they kept this relationship throughout the further month in which they were observed (Brown, 1954). On the other hand crabs flown through 51° longitude have been shown to retain the original phase-timing of the rhythm of colour change (Brown, Webb, and Bennett, 1955), despite the fact that any phasing factor dependent upon the rotation of the earth must have been delayed, relative to the crabs previous experience, by 3·3 h.

Similar studies on the effect of transporting organisms through a considerable number of degrees of latitude have been made, using the time-sense of the bee as an indicator process. Renner (1955) transported bees from Paris, where they had been trained to collect food from 8.15 to 10.15 p.m. French Daylight Time, to New York, by an overnight flight during which they were kept in constant light and temperature: while still in these environmental conditions the bees were seen to feed in New York 24 h after the time they had last fed in Paris, that is no correction had been made for local time. A different result however was obtained when bees trained in New York were transported to California and allowed to feed in the open air: in this case after transport the bees showed two clear maxima of searching activity, instead of the normal single peak. The time of the first maximum occurred $1\frac{1}{2}$ h later on the second day than on the first day, and on the second and third days the second maximum also moved to a later time. By the third day the time of the second maximum was closely related to the local time, the delay corresponding exactly to the difference in local time between the two environments: the time of the first maximum has been little affected by local factors and this maximum is thought by Renner to represent the expression of an endogenous rhythm (Renner, 1959).

The problem as to whether rhythms arise as a result of independent endogenous time-control, or through the reception of external time-signals will be discussed in the light of these results in the next chapter.

CHAPTER 8

# Discussion

THERE are two major barriers which stand in the way of an elucidation of the problem of the control of diurnal rhythms. One of these arises because rhythmical processes must at all times be affected by the physiological state of the organism, by the state of its receptor mechanisms, and by its behaviour patterns: in the course of a lifetime ageing may alter the physiological equilibrium, the sensitivity of receptor mechanisms may vary, environmental influences may have permanent but perhaps delayed effects, and, at least in animals, learning may influence the behaviour. As the status of these factors is unknown in any organism being tested it may be difficult to distinguish between these effects and the characteristics of the rhythms. The second barrier is our present inability to distinguish between the 'hands' or indicator processes and the 'clock mechanism'. Despite these formidable problems it is possible to speculate about the types of mechanism which might control the rhythmicity of living organisms largely because, as has been seen in the preceding chapters, certain characteristics can be recognized which are common to all rhythmical processes so far studied. These characteristics are summarized below.

We know that in a natural environment the timing of the phases of any rhythm is, within a pattern imposed by the genetical constitution of the organism, primarily affected by variation in the light conditions of the environment; the effective conditions including both the degree of intensity in both the day and the night period and the photofraction. Superimposed on the primary effects of the light cycle are variations induced by (a) fluctuating temperature and humidity, (b) the amount of food present and the type of substratum and surrounding of the environment, (c) the age, sex and physiological condition of the organism, and (d) the number of other animals of the same species present in the immediate environment, and the proportion of their sexes. Organisms can be divided fairly clearly, on the basis of their reactions to these variables, into the two major groups of diurnal and nocturnal organisms.

When an organism is removed from a fluctuating environment the phase-setting of its diurnal rhythms remains relatively stable, as does the period. In a constant environment the period of a rhythmical process appears to be determined by (a) the genetical constitution of the organism, (b) its physiological state, (c) the relationship between the phases of its several rhythmical processes, (d) the features of the constant environment in which it is being observed. The period is only slightly, if at all, affected by temperature. Whatever the state of these variables, and however they are combined at any one time, the range of periodicity in the large majority of organisms is confined to $24 \pm 2$ h.

Within any one 24-h period the degree of phase-shift which can be induced by a change in environmental conditions is limited to a few hours, seldom more than four. It is argued that this limitation in the degree of possible phase-shift is a reflection of the limitation in the period range which is imposed by the genetical constitution of the organism. A single change in environmental conditions may continue to influence the phase-setting of a rhythm for many days, and may result in a series of phase-shifts on successive days.

Phase-shifts can be induced by a limited range of metabolic inhibitors, but, as with the effect of low temperature, no permanent effect on the period of the organism can be induced except such as may also be induced by any environmental variable. It is however possible to inhibit the rhythm completely by affecting the RNA synthesis.

The phases of several rhythms within the one organism may become dissociated as a consequence of differential rates of phase-shifting, but no case of permanent dissociation has been recorded. There is evidence that experimental enforcement of dissociation is followed by physiological stress and pathological disorder.

Such then are the characteristics of diurnal rhythms, and it is in relation to these characteristics that any hypothesis concerning the controlling mechanism of the clock system must be examined.

The fundamental divergence between the hypothesis that rhythmicity is under the control of an internal, or endogenous, mechanism and the hypothesis that it is under the control of external time-signals which are perceived by the organism, has already been mentioned. The main criticism of the theory of external control is based on the fact that the phases of a rhythm can be set to lie in any relationship with solar time, and that in a constant environment the free-running period is only exceptionally exactly 24 h. The proponents of the theory of external control contend that the setting of the phases is analogous to moving the hands of a clock, and that this in no way affects the periodicity of the clock, and in reply to the second criticism suggest that an organism reacts to a constant environment by undergoing very small phase-shifts (autophasing) so that the periodicity appears to be other than 24 h. Autophasing is said to be due to the action of the environment on a cycle of sensitivity to either light or darkness. Since we know that the action of light, when it occurs during the subjective night of an organism, does indeed cause a phase-shift this seems a not unreasonable hypothesis. If a rhythm of sensitivity to the environment is produced by external time-signals presumably the phase-setting would be such that the light-sensitive period would coincide with solar night-time, and the dark-sensitive period with solar day-time. No difficulty is then encountered in explaining the 'drifting' of free-running rhythms, for in either constant darkness or light some part of the 24-h cycle would be affected by the environment and phase-shifting would follow. On the other hand it would be expected that when organisms living in constant darkness were exposed to light for brief periods they would show a differential response in terms of phase-shift at different times of solar day. However such differential phase-shift responses can only be measured relative to the time of subjective night and not to the time of solar night. That is, a period of light causes a similar phase-shift if it is given say 3 h after the peak of an animal's running activity, whether the running activity at the time occurs during solar day-time or solar night-time. This lack of response to light as measured in terms of phase-shift relative to solar time appears to be a considerable weakness in the case for exogenous control.

Another possible difficulty arises from the discovery that the period of a free-running rhythm is permanently altered after even a single exposure to light (p. 28); this, however, is not an insuperable difficulty since the exogenous control theory already assumes a specific degree of phase-shift by the organism when it is in constant conditions, and it could be argued that the phase-shifting mechanism is permanently affected by light. Indeed just such a complex mechanism must be postulated to explain the reaction if an endogenous timing system is assumed.

The case for the endogenous controlling system has been criticized by Brown mainly on the grounds that variables more stable than light or temperature are needed to maintain the accurate timing of organisms living in natural conditions, and that what is required are stable forces demanding little or no adaptive response, forces which are so 'pervasive that no living thing would ever be normally deprived of their influence'. It is true of course that the primary phase-setting factor, light, is relatively variable from day to day, but as has been seen other qualities of the light cycle, and the interaction of many other environmental variables may act together to prevent random fluctuations in light intensity from destroying the periodicity of a rhythm.

At the present time neither the internal clock theory nor the external time-signal theory can be proved, but even should the external time-signal theory be correct some internal mechanism must still be involved both for the reception of the signals and their transmission to the indicator processes. An internal mechanism must also be concerned with the maintenance of rhythms over short periods of time, for, as has been described (p. 89), organisms flown across many degrees of longitude do not phase-shift immediately in response to their new solar time, and are therefore maintaining a rhythm independently of external signals. The question of the position and mechanism of the internal timing system therefore arises.

It has been proposed (Harker, 1958a), and is now generally accepted, that the internal clock system in multicellular animals is not confined to one particular region of the body. There may be a basic metabolic circadian rhythm present in all cells of plants and animals, but different groups of cells may be affected by different environmental factors and each may constitute a 'physiological clock'; such clocks may in turn regulate the rhythm, or at least the phase-setting, of other groups of cells. The evidence on which this theory was based, when first proposed, was drawn largely from the fact that both unicellular organisms and separate tissues of multicellular organisms show circadian rhythms, facts which suggest that specialized mechanisms such as endocrine or nervous systems are not necessary for the maintenance of rhythms. The separation of at least two distinct clock systems, each of which can maintain an accurate periodicity in the absence of the other, has since been achieved in the cockroach. These two clock systems are in fact specialized systems, one being a neurosecretory system, the other is thought to be associated with the nervous system. On the other hand it has also been shown that the cells of the midgut show a rhythm of sensitivity to a hormone, and that this rhythm is independent of the known endocrine systems and probably of the nervous system (Harker, 1958b and unpublished). It is not unreasonable to suppose that the timing mechanisms of complex animals have undergone the type of specialization with which we are familiar in the parallel case of the nervous system. In the course of evolution the nervous system has become specialized so that in different parts of the body various concentrations of nervous tissue have taken on different functions, although the actual mechanism of nervous conduction remains the same in all cases. Similarly we may find that the basic mechanism of time measurement is the same for all animals (and plants), but that various tissues or organs have become specialized to play different parts in the complex of processes which control or produce the rhythms of highly specialized organisms.

However specialized the timing mechanism may have become in complex animals the basic control must still be a cellular one, it is therefore worth speculating about the possible methods of control at the cellular level.

The simplest type of indicator process would appear to be a single enzyme reaction, and there are three ways in which the rate of such a reaction could be rhythmically determined.

(1) There might be a cyclical variation in the concentration of the enzyme involved, regulated perhaps through a rhythm in the rate of protein synthesis.

(2) There might be a cyclical variation in the concentration of the substrate, this might come through variation in membrane permeability and therefore variation in flow of substrate into the reacting system.

94

(3) There might be rhythmical changes in the rate constants, brought about by some change in the microenvironment around the active centres, for example changes in pH or salt concentration producing a change in the protein configuration. This also might be related to changes in membrane permeability.

The only rhythmical process approaching this simplicity, about which any of the details are known, is the luminescent reaction in *Gonyaulax*. In this reaction purified extracts of the enzyme luciferase will only react with the substrate luciferin in the presence of salt (Hastings and Bode, 1961). The salt probably produces changes in the tertiary structure of the protein and hence in the catalytic activity. Within the cell therefore changing salt concentrations could be affecting the rhythm of luminescence, but we are still left with the three possible variations suggested above (1) a rhythm of substrate concentration, (2) a rhythm of total enzyme concentration, or (3) a rhythm of enzyme activity related to changes in the microscopic environment, for example the salt concentration. There are two basic processes which could control any of these, protein synthesis directly, or an indirect effect of protein synthesis on the structure and permeability of the intracellular membranes.

Very little is known about the role of protein synthesis in rhythmical processes but, again in *Gonyaulax*, it has been shown that the rhythm of luminescence is inhibited by actinomycin D, which, it is suggested, inhibits the DNA-dependent RNA synthesis. On the other hand chloramphenicol stimulates the amplitude of the rhythm (by stimulating the production of messenger-like RNA?). These two pieces of evidence suggest that the rhythm is dependent on cyclical variations in the level of messenger-RNA. A close connection between messenger-RNA, with or without the involvement of protein synthesis is an attractive hypothesis and seems to be the most promising line for further research. Furthermore the fact that two rhythms in *Gonyaulax*, that of luminescence and that of photosynthesis, appear to be differentially sensitive to RNA inhibitors suggests that this may be due to different types of RNA or different turnover times of the appropriate RNA or protein. This type of basic mechanism could be used to control any process and could be adapted in more complex organisms for the regulation of clock mechanisms such as those provided by neurosecretory cells.

95

Whether this hypothesis proves to be correct or not there seems to be, at the least, strong evidence that circadian rhythms are due to some type of cyclical variation in the state of the cell's macromolecules. We are still largely in the dark about many of the fundamental cellular processes. It may be that we need a better understanding of the details of nucleic acid replication, protein synthesis and the processes of energy transfer before further progress can be made. There is evidence that many of the highly organized macromolecular complexes can function as a single unit, as for example the sub-units of the mitochondria, and it may be that periodicity is a property which can be built into some such type of organization. The answer to the problem of the biological clock may lie in the realms of quantum biochemistry rather than in the more familiar terms of biology.

The problem of the mechanism of phase-setting is likely to remain obscure until more is known about the timing mechanism. It is interesting, in view of the possible implication of RNA in the timing mechanism, that in those unicellular organisms for which action spectra have been studied, phase-shifting is most effectively brought about by ultra-violet irradiation, and in the more complex cockroach X-irradiation also produces phase-shifting. Both of these instances suggest the implication of nucleic acid metabolism. The answer is not, however, likely to be a simple one, for not only does the phase-shift produced by ultra-violet occur in opposite directions in the two unicellular species studied, *Gonyaulax* and *Paramaecium*, but also it must be remembered that a number of other environmental variables can determine phase. One of the major difficulties in the elucidation of this problem is the separation of the intermediary reactions between the clock and the indicator process. It seems likely that where pigments or specialized receptor mechanisms are present these may be used to transmit information to the primary clock mechanism. For instance the action spectra of *Gonyaulax* shows a high efficiency at the 400–500 m$\mu$ wavelength, which may be related to the presence of chlorophyll, but this need not imply that chlorophyll is directly concerned in the clock mechanism. In the case of the cockroach there is evidence that phase-setting is primarily determined by the light-off signal, and that this is transmitted from the specialized receptors, the ocelli, via the nervous system to the neurosecretory cells of the sub-oesophageal ganglion. These neurosecretory cells in their turn act as a clock system by releasing hormonal material periodically and, as is to be expected, such hormonal rhythmicity produces a rhythmicity in a number of indicator processes.

In our present state of ignorance it is difficult to assess the significance of the temperature-independence of the period of the rhythm. Whereas we may reasonably expect the amplitude of the rhythm of an indicator process to show normal temperature-dependence (as indeed it does), it is not as clear how we would in fact expect the period to vary with temperature. Although the reciprocal of the period has the dimensions of a rate it is not clear how we should interpret this 'rate' in chemical terms. It may be not only the rate but also the extent of a process which is involved in the triggering of the indicator system, and both of these may be affected by temperature.

It is possible that the critical factor for the indicator process is a threshold level of some particular substance in a subcellular compartment, for example the ribosomes, some phase of the endoplasmic reticulum, or other particulate inclusion in the cytoplasm. The concentration will then be affected by two factors, the total amount of the substance, and the volume of the subcellular phase or the number of inclusions. Each of these might be affected by temperature in such a way that the concentration would be relatively little affected, and thus the threshold value would be reached in the same time. This example is purely speculative and is intended only to draw attention to the fact that temperature-independence of the period may not be as surprising as would seem at first sight.

A suggestion that the clock mechanism may function even when there is no observable rhythmicity in an indicator process has previously been made (Harker, 1958a). The evidence for this is twofold.

(1) A behaviour rhythm, such as running activity in the cockroach, may cease to be apparent for many days, yet may reappear at a time of solar day which indicates that an unchanged periodicity has been pursued by the controlling mechanism. This can be most clearly seen in fig. 38.

(2) Activities which take place only very infrequently, such as feeding of the bed bug *Cimex*, or the spawning of molluscs, show a fairly precise timing.

It seems likely that such interruptions in the rhythm of

7                                         97

indicator processes may be brought about, particularly in the case of the latter group of activities, by the overriding physiological processes lying between the clock and the indicator process. Even in the case of activities such as running, which actually continue although their rhythmicity ceases, it is possible that intermediary physiological processes intervene. For example it is known that the rhythmicity of the running activity of the cockroach is primarily dependent on the rhythmicity of the secretion from the neurosecretory cells of the sub-oesophageal ganglion, but that this is secondarily dependent on

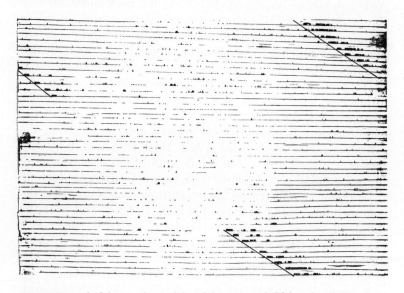

Fig. 38. Record of the running activity of the cockroach *Blaberus* in constant darkness, showing the lack of an observable rhythm for a period of 22 days and its reappearance at the time of day which would be expected if the rhythm had continued.

the continued secretion of a hormone which is transmitted from the corpus cardiacum to the sub-oesophageal ganglion. In the absence of the corpus cardiacum hormone the secretory activity of the neurosecretory cells continues for only a few days. Yet when this hormone is restored the timing of the renewed neurosecretory activity is such that it must be assumed that the cells continued to measure time throughout the non-secretory period.

It is further possible that interruptions in the rhythm of some indicator processes may be brought about because the critical factor for the indicator process is a threshold level of some particular substance (as has been discussed in relation to temperature-independence). Thus a rhythm reaching only sub-threshold levels of this substance will not affect the indicator process.

It is clear, from the evidence given in chapter 1, that rhythmical activities fall fairly distinctly into the two groups, nocturnal and diurnal. In one sense this may be a false impression since a great many processes within a single organism may be rhythmical, some reaching their peak in the light, others in the dark, but in most animals the major activity of locomotion, from which the timing of many others follow, seems to fall clearly into one or other group. The fact that animals are known to change from one group to the other in environments where competition is high suggests that the difference between the two types lies in the intermediary processes rather than in the basic clock mechanism.

Should the basic mechanism of the biological clock be discovered there will clearly remain very many problems concerning the way in which the information given by the cellular controlling system is translated into terms of the indicator process. In particular one problem stands out by its complexity, that is the problem of how animals use a timing mechanism as a chronometer or 'continuously consulted clock' during navigation. This subject has not been discussed in the previous chapters as it is too large a problem to consider in a book of this size. There is, however, convincing evidence that many animals, including those which travel only very short distances, for example within the confines of a pond, determine the direction of their movements by reference to celestial bodies. Since these bodies move relative to the earth a time factor must be involved in any type of navigation determined by their position. The properties of the time-sense involved in such navigational activities have been shown to be the same in all respects as those of the time mechanisms which have been described here.

The phenomenon of circadian rhythmicity, as has been shown, is widespread, and is found in many types of organism and in many processes of physiology and behaviour. Since Life evolved in an environment with a 24-h periodicity it may be that all living organisms consequently have an inherent natural frequency corresponding to this, but it appears that this basic rhythmicity has been adapted to give particular advantages to different species.

By separation of activities in time rather than in space two species may make more efficient use of the one habitat, and rhythmicity in behaviour can thus provide a way of decreasing inter-specific competition. If such rhythmicity of behaviour is present in a number of species in the same environment there are not only obvious advantages for competitors in having complementary phase-timing, but also for interdependent species in having carefully synchronized phase-timing. For example it is clearly mutually advantageous that the rhythm of opening of nocturnal entomophilous flowers is synchronized with the time of activity of their insect pollinators, and that the rhythm of feeding activity of bees can be modified to follow the rhythm of nectar secretion of the plants they visit.

The synchronization of the rhythms of individuals within the same species may also have advantages; this is particularly so in the case of sessile or semi-sessile marine forms which, because fertilization is external, shed sperms and eggs into the sea.

As an aid to survival the timing of activity in soft-bodied animals is also important; it may be a considerable advantage for an animal which normally moves out from its dark shelter only at night-time if it is able to time its excursions in some way other than by exposing part of its body in order to discover whether it is light or dark outside. Similarly an emergence or eclosion rhythm in an insect can be recognized as advantageous. For example a fly may leave the puparium and move upwards through several inches of soil, but once it has reached the surface it cannot remain in the soil until the right external conditions arise, for it is essential that the wings should be extended before the hardening of the cuticle is completed. As the animal cannot consult the external environment before eclosion, it is important that the onset of eclosion should be controlled internally in order to prevent the vulnerable adult, which is unhardened and as yet unable to fly, reaching the surface at an unsuitable time of day.

The adaptive significance of some rhythms of behaviour is not however very obvious, and indeed in some cases it would seem that an immediate reaction to the prevailing environment would have many advantages. It may be that our knowledge of the biology of those organisms which show such rhythms is insufficient for us to recognize the advantages, on the other hand it may be that indeed they do not have an adaptive significance but have arisen secondarily as a result of physiological rhythms which are also being pursued.

The importance of physiological rhythms, as distinct from those of behaviour, has been largely overlooked by biologists. Physiological regulation as a whole can be viewed in terms of supplying the right amount of material to the right place, but such regulation would be inefficient if the third specification—at the right time—were not fulfilled. Biological processes require finite time and it is often advantageous for an organism to be able to anticipate and prepare for predictable events. The possession of a time sense allows an organism to set in motion, at the proper time, processes whose final stage should be synchronized with set phases of a rhythmical environment.

# REFERENCES

AGREN, G. (1935). Die cyclischen Veränderungen im Leberglykogen von Ratten nach Nebennierenexstirpation. *Biochem. Z.* **281**, 367.

ANLICKER, J. and MAYER, J. (1956). An operant conditioning technique for studying fasting patterns in normal and obese mice. *J. appl. Physiol.* **8**, 667.

ASCHOFF, J. (1953). Aktivitätsperiodick von Mausen im Dauerdunkel. *Arch. ges. Physiol. Pflüg.* **255**, 189.

ASCHOFF, J. (1958). Tierische Periodik unter dem Einfluss von Zeitgebern. *Z. Tierpsychol.* **15**, 1.

ASCHOFF, J. (1960). Exogenous and endogenous components in circadian rhythms. *Cold Spring Harbor Symp. on Quant. Biol.* **25**, 11.

ASCHOFF, J. and MEYER-LOHMANN, J. (1955). Die Aktivitaetsperiodik von Nagern im kuenstlichen 24-stunden-Tag mit 6 bis 20 Stunden Lichzeit. *Z. vergl. Physiol.* **37**, 107.

BACQ, Z. M. (1929). Sur l'existence d'un rhythme nycthéméral de métabolisme chez le Coq. *Ann. Physiol.* **5**, 497.

BALDWIN, S. P. and KENDEIGH, S. C. (1932). Physiology of the temperature of birds. *Cleveland Mus. Nat. Hist. Sci. Publ.* **3**, 1.

BARE, J. K. (1959). Hunger, deprivation, and the day–night cycle. *J. comp. Physiol. Psychol.* **52**, 129.

BARNES, H. F. (1930). On some factors governing the emergence of gall-midges (Cecidomyidae). *Proc. zool. Soc. Lond.* p. 381.

BARROTT, H. G., FRITZ, J. C., PRINGLE, E. M. and TITUS, H. W. (1938). Heat production and gaseous metabolism of young male chicks. *J. Nutr.* **15**, 145.

BATEMAN, M. A. (1955). The effect of light and temperature on the rhythm of pupal ecdysis in the Queensland fruit-fly *Dacus tryoni*. *Aust. J. Zool.* **3**, 22.

BENTLEY, E. W., GUNN, D. L. and EWER, D. W. (1941). The biology and behaviour of *Ptinus tectus*, a pest of stored products. I. The daily rhythm of locomotor activity, especially in relation to light and temperature. *J. exp. Biol.* **18**, 182.

BLISS, D. E. (1962). Neuroendocrine control of locomotor activity in the land crab *Gecarcinus lateralis*. *Mem. soc. Endocrinology.* **12**. *Neurosecretion*, p. 391.

BODENHEIMER, F. S. and KLEIN, H. Z. (1930). Über die Temperatur-abhängigkeiten von Insekten. *Z. vergl. Physiol.* **11**, 345.

BOHN, G. and DRZEWINA, A. (1928). Les 'Convoluta'. *Ann. Sci. nat. (Zool.)*, **11**, 299.

BRETT, W. J. (1955). Persistent diurnal rhythmicity in *Drosophila* emergence. *Ann. ent. Soc. Amer.* **48**, 119.

BROWMAN, L. G. (1943). The effect of controlled temperatures upon the spontaneous activity rhythms of the albino rat. *J. exp. Zool.* **94**, 477.

Brown, F. A. (1954) Biological clocks and the fiddler crab. *Sci. Amer.* **190**, 34.

Brown, F. A. (1960). Response to pervasive geophysical factors and the biological clock problem. *Cold spring Harbor Symp. on Quant. Biol.* **25**, 57.

Brown, F. A., Bennett, M. F. and Webb, H. M. (1953). Endogenously regulated diurnal and tidal rhythms in metabolic rate in *Uca pugnax*. *Biol. Bull., Wood's Hole*, **105**, 371.

Brown, F. A., Bennett, M. F. and Webb, H. M. (1954). Persistent daily and tidal rhythms of oxygen consumption in the fiddler crab. *J. cell. comp. Physiol.* **44**, 477.

Brown, F. A., Bennett, M. F. and Webb, H. M. (1960). An organismic magnetic compass response. *Biol. Bull., Wood's Hole,* **119**, 65.

Brown, F. A., Brett, W. J., Bennett, M. F. and Barnwell, F. H. (1960). Magnetic response of an organism and its solar relationships. *Biol. Bull., Wood's Hole,* **118**, 367.

Brown, F. A., Fingerman, M. and Hines, M. N. (1954). A study of the mechanism involved in shifting of the phases of the endogenous daily rhythm by light stimulus. *Biol. Bull., Wood's Hole,* **106**, 308.

Brown, F. A., Freeland, M. and Ralph, C. L. (1955). Persistent rhythm of oxygen consumption in potatoes, carrots, and a sea-weed *Fucus*. *Plant Physiol.* **30**, 280.

Brown, F. A., Hines, M. N., Webb, H. M. and Fingerman, M. (1950). Effects of constant illumination upon the magnitude of the diurnal rhythm of *Uca*. *Anat. Rec.* **108**, 3.

Brown, F. A., Sandeen, M. I. and Fingerman, M. (1952). Modification of the tidal rhythm of *Uca* by tidal differences and by illumination. *Biol. Bull., Wood's Hole,* **103**, 397.

Brown, F. A. and Webb, H. M. (1948). Temperature relations of an endogenous daily rhythmicity in the fiddler crab, *Uca*. *Physiol. Zoöl.* **21** 371.

Brown, F. A., Webb, H. M. and Bennett, M. F. (1955). Proof for the endogenous component in persistent solar and lunar rhythmicity in organisms. *Proc. nat. Acad. Sci., Wash.,* **41**, 93.

Brown, F. A., Webb, H. M. and Bennett, M. F. (1958). Comparison of some fluctuations in cosmic radiation and organismic activity during 1954, 1955 and 1956. *Amer. J. Physiol.* **195**, 237.

Brown, F. A., Webb, H. M., Bennett, M. F. and Sandeen, M. I. (1955). Evidence for an exogenous contribution to persistent diurnal and lunar rhythmicity under so-called constant conditions. *Biol. Bull., Wood's Hole,* **109**, 238.

Brown, F. A., Webb, H. M. and Brett, W. J. (1960). Magnetic response of an organism and its lunar relationships. *Biol. Bull., Wood's Hole,* **118**, 382.

Brown, F. A., Webb, H. M. and Graves, R. C. (1952). A persistent tidal rhythm in the fiddler crab *Uca pugnax*. *Biol. Bull., Wood's Hole,* **103**, 297.

Brown, R. H. J. and Harker, J. E. (1960). A method of controlling the temperature of insect neurosecretory cells *in situ*. *Nature, Lond.,* **185**, 392.

BRUCE, V. G. (1960). Environmental entrainment of circadian rhythms. *Cold Spring Harbor Symp. on Quant. Biol.* **25**, 29.

BRUCE, V. G. and PITTENDRIGH, C. S. (1956). Temperature independence in a unicellular 'clock'. *Proc. nat. Acad. Sci., Wash.*, **42**, 676.

BRUCE, V. G. and PITTENDRIGH, C. S. (1960). An effect of heavy water on the phase and period of the circadian rhythm in *Euglena. J. cell. comp. Physiol.* **56**, 25.

BÜHNEMANN, F. (1955 *a*). Das endodiurnale System der Oedogoniumzelle. IV. Die Wirkung verschiedener spektralbereiche auf die Sporulation und Mitoserhythmik. *Planta*, **46**, 227.

BÜHNEMANN, F. (1955 *b*). Das endodiurnale System der Oedogoniumzelle. III. *Zeit. Naturforsch.* **10b**, 305.

BULLOUGH, W. S. (1962). The control of mitotic activity in adult mammalian tissues. *Biol. Rev.* **37**, 307.

BULLOUGH, W. S. and LAURENCE, E. B. (1961). Stress and adrenaline in relation to the diurnal cycle of epidermal mitotic activity in adult male mice. *Proc. roy. Soc.* B. **154**, 540.

BUNNING, E. (1958). Uber den Temperatureinfluss auf die endogene Tagesrhythmik besonders bei *Periplaneta americana. Biol. Zentralbl.* **77**, 141.

BUNNING, E. (1959). Physiological mechanisms and biological importance of the endogenous diurnal periodicity in plants and animals. *Photoperiodism and Related Phenomena in Plants and Animals*, ed. Withrow, p. 507. Washington: A.A.A.S.

BUNNING, E. and SCHÖNE-SCHNEIDERHÖHN, G. (1957). Die Bedeutung der Zellkerne im Mechanismus der endogenen Tagesrhthmik. *Planta*, **48**, 459.

BURCKARD, E., DONTCHEFF, L. and KAYSER, C. (1933). Le rhythme nycthéméral chez le Pigeon. *Ann. Physiol.* **9**, 303.

CALHOUN, J. B. (1944). Twenty-four hour periodicity in the animal kingdom. *J. Tenn. Acad. Sci.* **19**, 179.

CASPERS, H. (1953). Rhythmische Erscheinungen in der Fortpflanzung von *Clunio marinus* und das Problem der lunaren Periodizitat bei Organismen. *Arch. Hydrobiol.* **18** (Suppl.), 415.

CHITTY, D. and SOUTHERN, H. N. (1954). *Control of Rats and Mice.* Oxford.

CLOUDSLEY-THOMPSON, J. L. (1956). Studies in diurnal rhythms. VI. *Ann. Mag. nat. Hist.* **9**, 305.

COLD SPRING HARBOR SYMPOSIA ON QUANTITATIVE BIOLOGY (1960). Volume XXV: *Biological clocks.*

COLE, L. C. (1957). Biological clock in the unicorn. *Science*, **125**, 874.

CORBET, P. S. (1957). The life-history of the Emperor Dragonfly, *Anax imperator* Leach. *J. Anim. Ecol.* **26**, 1.

CORBET, P. S. (1960). Patterns of circadian rhythms in insects. *Cold Spring Harbor Symp. on Quant. Biol.* **25**, 357.

DECOURSEY, P. (1960). Phase control of activity in a rodent. *Cold Spring Harbor Symp. on Quant. Biol.* **25**, 49.

DELL, P. (1961). In *Ciba Foundation Symp. on 'The Nature of Sleep,'* p. 391. Churchill, London.

DE MAIRAN (1729). Observation botanique. *Histoire de l'Academie Royale des Sciences*, p. 34. Paris.

EHRET, C. F. (1959). Induction of phase shift in cellular rhythmicity by far ultraviolet and its restoration by visible radiant energy. *Photoperiodism and Related Phenomena in Plants and Animals*, ed. Withrow, p. 541. Washington: A.A.A.S.

EHRET, C. F. (1960). Action spectra and nucleic acid metabolism in circadian rhythms at the cellular level. *Cold Spring Harbor Symp. on Quant. Biol.* **25**, 149.

EIDMANN, H. (1955). Ueber rhythmische Erscheinungen bei der Stabhleuschrecke *Carausius morosus* Br. *Z. vergl. Physiol.* **38**, 370.

EVERETT, J. W. and SAWYER, C. H. (1950). A 24-hour periodicity in the 'L-H-release apparatus' of female rats disclosed by barbitol sedation. *Endocrinology*, **47**, 198.

EWER, D. W. and EWER, R. F. (1941). The biology and behaviour of *Ptinus tectus* Boie., a pest of stored products. *J. exp. Biol.* **18**, 290.

FLEESON, W., GLUECK, B. C. and HALBERG, F. (1957). Persistence of daily rhythms in eosinophil count and rectal temperature during 'regression' induced by intensive electroshock therapy. *Physiologist*, **1**, 28.

FOLK, G. E. (1957). Twenty-four hour rhythms of mammals in a cold environment. *Amer. nat.* **xci**, 153.

FRAPS, R. M. (1954). Neural basis of diurnal periodicity in release of ovule-inducing hormone in the fowl. *Proc. nat. Acad. Sci., Wash.*, **40**, 348.

FRENCH, J. D., VERSEANO, M. and MAGOUN, H. W. (1953). A neural basis of the anesthetic state. *Am. Med. Assoc. Arch. Neurol. Psychiat.* **69**, 519.

GRIFFIN, D. T. and WELSH, J. H. (1937). Activity rhythms in bats under constant external conditions. *J. Mammal.* **18**, 337.

HADDOW, A. J. (1961). Entomological studies from a high tower in Mpanga Forest, Uganda. *Trans. R. ent. Soc., Lond.*, **116**, 315.

HADDOW, A. J. and GILLETT, J. D. (1957). Observations on the oviposition-cycle of *Aedes aegypti*. *Ann. trop. Med. Parasit.* **51**, 159.

HADDOW, A. J. and GILLETT, J. D. (1958). Laboratory observations on the oviposition-cycle in the mosquito *Taeniorhynchus (Coquillettidia) fuscopennatus* Theobald. *Ann. trop. Med. Parasit.* **52**, 320.

HALBERG, F. (1957). Young NH-mice for the study of mitosis in intact livers. *Experientia*, **13**, 502.

HALBERG, F. (1959). Physiologic 24-hour periodicity; general and procedural considerations with reference to the adrenal cycle. *Z. Vitam.- Horm.- u. Fermentforsch.* **10**, 225.

HALBERG, F., HALBERG, E., BARNUM, C. P. and BITTNER, J. J. (1959). Physiologic 24-hour periodicity in human beings and mice, the lighting regimen and daily routine. *Photoperiodism and Related Phenomena in Plants and Animals*, ed. Withrow, p. 803. Washington: A.A.A.S.

HALBERG, F., PETERSON, R. E. and SILBER, R. H. (1959). Phase relations of 24-hour periodicities in blood corticosterone, mitosis in cortical adrenal parenchyma, and total body activity. *Endocrinology*, **64**, 222.

HALBERG, F., ZANDER, H. A., HOUGHLAM, M. W. and MUHLEMANN, H. R. (1954). Daily variations in tissue mitosis, blood eosinophils and rectal temperatures of rats. *Amer. J. Physiol.* **177**, 361.

HARKER, J. E. (1953). The diurnal rhythm of activity of mayfly nymphs. *J. exp. Biol.* **30**, 525.

HARKER, J. E. (1954). Diurnal rhythms in *Periplaneta americana* L. *Nature, Lond.*, **173**, 689.

HARKER, J. E. (1955). Control of diurnal rhythms of activity in *Periplaneta americana* L. *Nature, Lond.*, **175**, 733.

HARKER, J. E. (1956). Factors controlling the diurnal rhythm of activity of *Periplaneta americana* L. *J. exp. Biol.* **33**, 224.

HARKER, J. E. (1958a). Diurnal rhythms in the animal kingdom. *Biol. Rev.* **33**, 1.

HARKER, J. E. (1958b). Experimental production of midgut tumours in *Periplaneta americana* L. *J. exp. Biol.* **35**, 251.

HARKER, J. E. (1960a). The effect of perturbations in the environmental cycle on the diurnal rhythm of activity of *Periplaneta americana* L. *J. exp. Biol.* **37**, 154.

HARKER, J. E. (1960b). Internal factors controlling the sub-oesophageal ganglion neurosecretory cycle in *Periplaneta americana* L. *J. exp. Biol.* **37**, 164.

HARKER, J. E. (1960c). Endocrine and nervous factors in insect circadian rhythms. *Cold Spring Harbor Symp. on Quant. Biol.* **25**, 279.

HARKER, J. E. (1961). Diurnal rhythms. *Ann. rev. Entom.* **6**, 131.

HARKER, J. E. (1963). Tumors. In *Insect Pathology. An Advanced Treatise*, vol. 1, p. 191. Academic Press.

HARKER, J. E. (1964a). The effect of a biological clock on the developmental rate of *Drosophila* pupae. In press.

HARKER, J. E. (1964b). Diurnal rhythms and homeostatic mechanisms. *Symp. soc. exp. Biol.* XVIII. In press.

HASTINGS, J. W. (1960). Biochemical aspects of rhythms: phase shifting by chemicals. *Cold Spring Harbor Symp. on Quant. Biol.* **25**, 131.

HASTINGS, J. W. and BODE, V. C. (1961). Ionic effects upon bioluminescence in *Gonyaulax* extracts. In *Light and Life*, ed. McElroy and Glass, p. 294. Baltimore.

HASTINGS, J. W. and SWEENEY, B. M. (1957a). On the mechanism of temperature independence in a biological clock. *Proc. nat. Acad. Sci., Wash.*, **43**, 804.

HASTINGS, J. W. and SWEENEY, B. M. (1957b). The luminescent reaction in extracts of the marine dinoflagelate *Gonyaulax polyedra*. *J. cell. comp. Physiol.* **49**, 209.

HASTINGS, J. W. and SWEENEY, B. M. (1958). A persistent diurnal rhythm of luminescence in *Gonyaulax polyedra*. *Biol. Bull., Wood's Hole*, **115**, 440.

HASTINGS, J. W. and SWEENEY, B. M. (1959). The *Gonyaulax* clock. *Photoperiodism and Related Phenomena in Plants and Animals*, ed. Withrow, p. 567. Washington: A.A.A.S.

HASTINGS, J. W. and SWEENEY, B. M. (1960). The action spectrum for shifting the phase of the rhythm of luminescence in *Gonyaulax polyedra*. *J. gen. Physiol.* **43**, 697.

HECKROTTE, C. (1962). The effect of the environmental factors on the locomotory activity of the Plains Garter Snake (*Thamnophis radix radix*). *An. Behaviour*, **10**, 193.

HEIBEL, G. (1949). Le rhythme nycthéméral de l'activité et de la calorification chez l'embryon de poulet et le jeune poulet (Light Sussex). *C. R. Soc. Biol.*, *Paris*, **143**, 864.

HELLBRUGGE, T. (1960). The development of circadian rhythms in infants. *Cold Spring Harbor Symp. on Quant. Biol.* **25**, 311.

HOFFMAN, K. (1957). Angeborene Tagesperiodik bei Eidechsen. *Naturwissenchaften*, **12**, 359.

HOFFMAN, K. (1960). Versuche zur Analyse der Tagesperiodik. I. *Z. vergl. Physiol.* **43**, 544.

HOWARD, R. B. (1952). Studies on the metabolism of iron. PhD. thesis, University of Minnesota.

HUECK, H. J. (1951). Influence of light upon the hatching of winter-eggs of the fruit tree red spider. *Nature, Lond.*, **167**, 993.

JOHNSON, M. (1939). Effect of continuous light on periodic spontaneous activity of white-footed mice (*Peromyscus*). *J. exp. Zool.* **82**, 315.

JUNDELL, J. (1904). Über die nykthemeralen Schwankungen im I. Lebensjahre des Menschen. *Jb. Kinderhk.* **59**, 521

KALMUS, H. (1934). Über die Natur des Zeitgedächtnisses der Bienen. *Z. vergl. Physiol.* **20**, 405.

KALMUS, H. (1955). Genetical responses to season and day. *Acta med. scand.* **307** suppl., 59.

KARAKASHIAN, M. W. and HASTINGS, J. W. (1962). The inhibition of a biological clock by Actinomycin D. *Proc. nat. Acad. Sci., Wash.*, **48**, 2130.

KAVANAU, I. L. (1962). Twilight transitions and biological rhythmicity. *Nature, Lond.*, **194**, 1293.

KLEITMAN, N. (1940). The modifiability of the diurnal pigmentary rhythm in isopods. *Biol. Bull., Wood's Hole*, **78**, 403.

KLEITMAN, N. (1952). Modifiability of the diurnal body temperature and heart rate. *Fed. Proc.* **11**, 83.

KLEITMAN, N. and ENGELMANN, T. G. (1953). Sleep characteristics of infants. *J. appl. Physiol.* **6**, 269.

KLUG, H. (1958). Histo-physiologische Untersuchungen über die Aktivitätsperiodik bei Carabiden. *Wiss. Z. Humbolt-Univ., Berlin*, **8**, 450.

KRAMER, G. (1952). Experiments on bird orientation. *Ibis*, **94**, 265.

LEVINSON, L., WELSH, J. H. and ABRAMOWITZ, A. A. (1941). Effect of hypophysectomy on diurnal rhythm of spontaneous activity in the rat. *Endocrinology*, **29**, 41.

LEWIS, P. R. and LOBBAN, M. C. (1957a). The effects of prolonged periods of life on abnormal time routines upon excretory rhythms in human subjects. *Quart. J. exp. Physiol.* **42**, 356.

LEWIS, P. R. and LOBBAN, M. C. (1957b). Dissociation of diurnal rhythms in human subjects living on abnormal time routines. *Quart. J. exp. Physiol.* **42**, 371.

MAGOUN, H. W. (1952). The ascending reticular activity system. *Proc. Ass. Res. nerv. Dis.* **30**, 480.

MORI, S. (1944). Daily frequency of activities of *Cavernularia obesa valenciennes*. III. Effect of light intensity. *Zool. Mag., Tokyo*, **56**, 1.

MORI, S. (1954). Population effect on the daily periodic emergence of *Drosophila*. *Mem. Coll. Sci. Kyoto*, (B), **25**, 49.

MORUZZI, G. (1961). In *Ciba Foundation Symp, on 'The Nature of Sleep,'* p. 393. Churchill, London.

MUNN, N. L. (1950). Handbook of psychological research on the rat. Harrap, London.

NIELSON, E. T. and HAEGER, J. S. (1954). Pupation and emergence in *Aedes taeniorhynchus* (Wied.). *Bull. ent. Res.* **45**, 757.

OHSAWA, W., MATUTANI, K., TUKUDA, H., MORI, S., MIYADI, D., YANAGISIMA, S. and SATO, Y. (1942). Sexual properties of the daily rhythmical activity in *Drosophila melanogaster. Physiol. Ecol.* **5**, 26.

PARK, O. (1935). Studies in nocturnal ecology. III. *Ecology,* **16**, 152.

PAULI, W. F. (1926). Versuche über dem physiologischen Farbenwechsel der Salamander Larve und der Pfrille. *Z. wiss. Zool.* **128**, 421.

PFEFFER, W. (1915). Beitraege zur Kenntnis der Entstehung der Schlafbewegungen. *Abhandl. saechs. Akad. wiss. math-physik.* **34**, 1.

PIRSON, A. and LORENZEN, H. (1958). Ein endogener Zeitfaktor bei Teilung von *Chlorella. Z. Bot.* **46**, 53.

PITTENDRIGH, C. S. (1954). On the temperature independence in the clock system controlling emergence time in *Drosophila. Proc. nat. Acad. Sci., Wash.* **40**, 1018.

PITTENDRIGH, C. S. (1958). Perspectives in the study of biological clocks. *Symp. on Perspectives in Marine Biol.,* p. 239. Berkeley.

PITTENDRIGH, C. S. (1959). Daily rhythms as coupled oscillator systems and their relation to thermoperiodism and photoperiodism. *Photoperiodism and Related Phenomena in Plants and Animals,* ed. Withrow. Washington: A.A.A.S.

PITTENDRIGH, C. S. (1960). Circadian rhythms and the circadian organization of living systems. *Cold Spring Harbor Symp. on Quant. Biol.* **25**, 159.

PITTENDRIGH, C. S. and BRUCE, V. G. (1957a). An oscillator model for biological clocks. *Rhythmic and Synthetic processes in Growth,* ed. Rudnick, p. 75. Princeton Univ. Press.

PITTENDRIGH, C. S. and BRUCE, V. G. (1957b). Endogenous rhythms in insects and microorganisms. *Amer. Nat.* **91**, 179.

PITTENDRIGH, C. S., BRUCE, V. G. and KAUS, P. (1958). On the significance of transients in daily rhythms. *Proc. nat. Acad. Sci., Wash.,* **44**, 965.

POHL, R. (1948). Tagesrhythmik im phototaktischen Verhalten der *Euglena gracilis. Z. Naturf.* **3b**, 367.

RALPH, C. L. (1959). Modifications of activity rhythms of *Periplaneta americana* L. induced by carbon dioxide and nitrogen. *Physiol. Zoöl.* **32**, 57.

RAWSON, K. S. (1959). Experimental modifications of mammalian endogenous activity rhythms. *Photoperiodism and Related Phenomena in Plants and Animals,* ed. Withrow. Washington: A.A.A.S.

RAWSON, K. S. (1960). Effects of tissue temperature on mammalian activity rhythms. *Cold Spring Harbor Symp. on Quant. Biol.* **25**, 105.

REMMERT, H. (1955). Untersuchungen über das tageszeitlich gebundene Schlüpfen von *Pseudosmittia arenaria. Z. vergl. Physiol.* **37**, 338.

RENNER, M. (1955). Ein Transozeanversuch zum Zeitsinn der Honigbiene. *Naturwissenschaften,* **42**, 540.

RENNER, M. (1959). The clock of the bees. *Nat. Hist. Mag.* **68**, 434.

RICHTER, C. P. (1927). Animal behaviour and internal drives. *Quart. Rev. Biol.* **2**, 307.

RICHTER, C. P. (1960). Biological clocks in medicine and psychiatry: shock-phase hypothesis. *Proc. nat. Acad. Sci., Wash.*, **46**, 1506.

ROBERTS, S. K. (1959). Circadian activity rhythms in cockroaches. Ph.D. thesis. Princeton University.

ROBERTS, S. K. (1962). Circadian activity rhythms in cockroaches. II. Entrainment and phase-shifting. *J. cell. comp. Physiol.* **59**, 175.

ROTHMANN, H. (1923). Zusammenfassender Bericht über den Rothmann-schen grosshirnlosen Hund nach klinischer und Anatomischer Untersuchung. *Z. ges. Neurol. Psychiat.* **87**, 247.

RUSSELL, G. V. (1957). The brainstem reticular formation. *Tex. Rep. Biol. Med.* **15**, 332.

SCOTT, W. N. (1936). An experimental analysis of the factors governing the hour of emergence of adult insects from the pupa. *Trans. R. ent. Soc. Lond.* **85**, 303.

SERFATY, A. (1945). Caractére du rhythme nycthéméral des larves d'Aeschnes. *Bull. Mus. Hist. nat., Paris*, **17**, 176.

SIEGEL, P. S. and STUCKEY, H. L. (1947). The diurnal course of water and food intake in the normal mature rat. *J. comp. Physiol. Psychol.* **40**, 365.

SJOEGREN, B., NORDENSKJOELD, T., HOLMGREN, H. and MOLLERSTRÖM, J. (1938). Beitrag zur Kettnis der Leberrhythmik. *Arch. ges. Physiol. Pflüg.* **240**, 427.

SKINNER, B. F. (1957). The experimental analysis of behaviour. *Amer. Sci.* **45**, 343.

SOLLBERGER, A. (1955a). Statistical aspects of diurnal biorhythm. *Acta Anat.* **22**, 97.

SOLLBERGER, A. (1955b). Diurnal changes in biological variability. *Acta Anat.* **23**, 259.

STARTZL, T. E., TAYLOR, C. W. and MAGOUN, H. W. (1951). Collateral excitation of reticular formation of brain stem. *J. Neurophysiol.* **14**, 479.

STEIN-BELING, I. (1935). Über das Zeitgedächtnis bei Tieren. *Biol. Rev.* **10**, 18.

STEINIGER, F. (1936). Die Biologie der sog 'tierischen Hypnose'. *Ergebn. Biol.* **13**, 348.

STEPHENS, G. C. (1957a). Responses of diurnal melanophore rhythms of *Uca pugnax* to changes in temperature. *Biol. Bull., Wood's Hole*, **109**, 352.

STEPHENS, G. C. (1957b). Influence of the rate and magnitude of temperature change on the diurnal melanophore rhythm of the fiddler crab *Uca pugnax. Biol. Bull., Wood's Hole*, **109**, 369.

STIER, J. B. (1930). Spontaneous activity of mice. *J. gen. Psychol.* **4**, 67.

STIER, J. B. (1933). On the temperature-regulating function of spontaneous activity in the mouse. *Proc. nat. Acad. Sci., Wash.*, **19**, 725.

SUOMALAINEN, P. (1961). Hibernation and sleep. In *Ciba Foundation Symp. on 'The Nature of Sleep'*, p. 307. Churchill, London

SWEENEY, B. M. and HASTINGS, J. W. (1958). Rhythmic cell division in populations of *Gonyaulax polyedra. J. Protozool.* **5**, 217.

SWEENEY, B. M. and HASTINGS, J. W. (1960). Effects of temperature upon diurnal rhythms. *Cold Spring Harbor Symp. on Quant. Biol.* **25**, 87.

Sweeney, B. M. and Haxo, F. (1961). Persistence of a photosynthetic rhythm in enucleated *Acetabularia. Science*, **134**, 1361.

Tjønneland, A. (1960). The flight activity of mayflies as expressed in some East African species. *Univ. Bergen Årb. naturv. R.* **1**, 1.

Tribukait, B. (1954). Aktivitätsperiodik der Maus im kunstlichen verkurtzten Tag. *Naturwissenschaften*, **41**, 92.

Tribukait, B. (1956). Die Aktivitätsperiodik der weissen Maus im Kunsttag von 16–29 Stunden Lange. *Z. vergl. Physiol.* **38**, 479.

Übelmesser, E. R. (1954). Über den endonomen Rhythmus der Sporangientrager bildung von *Pilobolus. Arch. Mikrobiol.* **20**, 1.

Verplanck, W. S. and Hayes, J. R. (1953). Eating and drinking as a function of maintenance of schedule. *J. comp. Physiol. Psychol.* **46**, 327.

Wahl, O. (1932). Neue Untersuchungen uber das Zeitgedächtnis der Bienen. *Z. vergl. Physiol.* **16**, 529.

Webb, H. M., Brown, F. A., Bennett, M. F., Shriner, J. and Brown, R. A. (1956). Similarities between daily fluctuations in the background radiation and oxygen consumption in the living organism. *Anat. Rec.* **125**, 615.

Webb, H. M., Brown, F. A. and Sandeen, M. I. (1954). A modification in the frequency of the persistent daily rhythm of the fiddler crab. *Anat. Rec.* **120**, 796.

Williams, C. B. (1939). An analysis of four year captures of insects in a light trap. *Trans. R. ent. Soc. Lond.* **89**, 79.

Williams, G. (1959). Seasonal and diurnal activity of Carabidae, with particular reference to *Nebria, Notiophilus* and *Feronia. J. Anim. Ecol.* **28**, 309.

Wolf, E. (1930). Die Aktivität der japonischen Tanzmaus und ihre rhythmische Verteilung. *Z. vergl. Physiol.* **11**, 321.

# INDEX

actinomycin, 60–1, 95
action spectra, 56–7
adrenalectomy, 73
adrenal gland, 60, 72–4
advance reset, 42–6
age and rhythms, 17, 76, 90
amphibia, 17, 84
amplitude, 41
  abnormal increase in, 82
  definition, 3
  temperature dependence, 97
anaesthetics, 53–4
ants
  activity, 16
  feeding rhythms, 13
arrhythmic activity, 11, 62
Aschoff's Rule, 22
autophasing, 92

barometric pressure, 15, 84–7
bats, 13, 26, 31
bees
  feeding rhythms, 13, 24, 100
  temperature effects, 23, 24
  transportation experiments, 89
beetles, 12, 16, 17, 22, 33
biological clock, types of, 2
birds (*see also* chaffinch)
  embryos, 76
  ovulation, 76
*Blaberus*, *see* cockroach

*Carausius* (stick insect), 22
carbohydrate metabolism, 74
carbon dioxide, 53
*Cavernularia* (sea pen), 9
cell division (*see also* Mitosis), 10
cellular rhythms, 59–62, 71, 93–6
chaffinches, 6, 22
chironomids, 22, 33
chloramphenicol, 60–1, 95
*Chlorella*, 10
*Cimex* (bed bug), 97
circadian rhythm, definition, 3

cockroach
  carbon dioxide effect, 54
  endocrine system, 62–72
  environmental perturbation, 38, 40
  food, 13
  light intensity effects, 22
  locomotor activity, 3, 11, 64
  nervous system, 70–1
  nitrogen effect, 54
  period length, 33, 35
  phase-setting, 7, 64
  phase-shift, 40, 41, 43–5, 52, 54
  previous environment, 27
  temperature effects, 12, 52
  transients, 40
coelenterates, *see Cavernularia*
*Convoluta*, 22
corpus allatum, 70, 98
corpus cardiacum, 70, 98
corticosterone, 72
cosmic radiation, 87
crabs (*see also Uca*), 72
crustacea, *see* crabs, *Uca*, *Ligia*.

*Dacus tryoni* (fruit fly), 12
delay reset, 42–6
deuterated water, 56
disphasia, 78
dissociation of rhythms, 31, 35, 78, 80, 91
diurnal animals
  and period, 20
  phase-setting, 8, 10, 99
DNA, 95
  relative specific activity, 59, 60
  synthesis, 55
dragonfly, 17, 22
*Drosophila*
  and cosmic radiation, 88
  developmental rates, 46–50
  eclosion rhythm, 10, 11, 46–50
  flight activity, 16
  nitrogen effect, 53
  phase-shift, 49–50, 53
  population effect, 17, 47, 50
  pupae, 47–50